A CONSULTANT'S STORY

A CONSULTANT'S STORY

How an Engineering Business Works Domestically and Overseas, with a focus on India

RICHARD MOORE

COVER PHOTO

See Blog entry, Saturday, February 25, 2012, Page 274

Printed in the United States of America

First Printing, 2026

ISBN: Trade Paperback: 978-82-18-92228-4
ISBN: eBook: 978-82-18-92229-0

Book Design by Glen M. Edelstein, Hudson Valley Book Design

Erin with her dog Winston

This book is dedicated to my daughter Erin, who not only edited my draft, but has been my shinning light through most of the time described in the following pages. She is beautiful, smart and always sure to tell me when I get off track. I'm proud of her, especially her efforts to get out of her comfort zone and explore the whole world.

CONTENTS

CHAPTER 1

INTRODUCTION

This is the third and final book in my memoir trilogy. The first, "A Charmed Young Life" covers the time from my birth in 1947 until I graduated from college in 1969. The second, "An Officer's Journey" covers the next two years that I spent in the Army, 1969-71. This third book covers the time after the Army, until I officially retired from full time work at age 65 in 2012.

I initially wanted this book to be about my time in India, at the end of my career. I started to get interested in India in 2008, but, except for several trips, did the work from the US. It became clear that the only way to make a difference was to go and stay for extended periods. So that is what I did from 2010-2012. During that time I kept a blog and I planned to use it as a reference while writing a new narrative. That got too complicated, so the blogs are presented in their original format, with minor edits and an introduction.

I also felt that I needed to say something about the 40 years between 1971 and my time in India. That effort got a bit out of hand and now forms the first half of this book. It's a good reference on how to be a consulting engineer. Within my own story you can see how a career is built by learning your profession, doing the nitty-gritty technical stuff, working with people (teams), participating in professional organizations and eventually leading. Leading is much more than the technical stuff and includes understanding people and the business side of the profession.

There are no shortcuts. You need to build a good foundation, so when you get an opportunity you have the tools and experience to succeed. It's also important to know that success doesn't always come the first time or the second. You will make many mistakes, but if you learn from them you will be a better leader in the long run.

If the book is too much to get through, I suggest you listen to the lecture I gave to Lafayette College students on the consulting business, which also touches on my time in India . Consider it the CliffsNotes version of this book. See Youtube link: https://youtu.be/OFbHjv6KQFw

So let's begin.

CHAPTER 2

UNIVERSITY OF MARYLAND, 1971-1972

After I got out of the Army I went to graduate school in the Civil Engineering Department at the University of Maryland on their main campus in College Park. This was just outside the District of Columbia, on the northern (Maryland) side, as opposed to my stay in the Army a year before in Virginia, on the southern side of the District. Living in Metro DC felt a bit like coming home, after Vietnam, and I rented an apartment in Adelphi MD not far from campus.

I had a simple first floor, one bedroom apartment, a short 15 minutes from Campus. I could afford the apartment for two important reasons. First, I was eligible for the GI Bill which amounted to about $3,000 to 4,000 dollars a year. I was also supported by a grant from the State of Maryland in approximately the same amount. Combined, I was making about $7,000/year, not a bad salary in 1971. One of my classmates, who was not on the GI Bill, thought I was rich.

I needed a master's degree in environmental engineering because my college courses focused on structural engineering and, despite my exposure to water and wastewater issues while teaching at Fort Belvoir, I had no formal education in the environmental area. At the time, 1971, environmental engineering was just getting up a head of steam because of all the headlines about water pollution: think of the Cuyahoga River catching on fire in Cleveland OH. A year later, the Clean Water Act was passed, with a boatload of federal money to address the problem. So my timing was perfect!

Environmental engineering can include a lot of things related to air, water and land, but for me it always focused on water. This includes potable (drinking) water, wastewater (sewage), storm water and water resource management. But like most things, you follow the money and in 1971, that included cleaning up dirty wastewater discharges from municipal and industrial facilities (called sanitary engineering at the time) and water resources. There was no federal money for potable (drinking) water supplies, so professors in the Civil Engineering Department applied for grants to fund their work in sanitary engineering and water resources. Most money went to planning, designing and constructing municipal wastewater collection systems (sewers) and treatment plants.

My advisor, Dr. Ragan, was head of the department and had a PhD in water resources from Cornell. His area of expertise was hydrology, especially the rainfall-runoff process. As such, he had a large grant from the state of Maryland to evaluate the Anacostia River watershed. It's a small river in Maryland, rising in the hilly piedmont, flowing into the flat coastal plain, just north of the District of Columbia. The river crosses into the District where it becomes tidal before joining the Potomac River, south of the city. I was paid with grant money, as mentioned above, to help Dr. Ragan. My assignment was to study urbanization in the watershed and evaluate its impact on hydrology, erosion and sedimentation. This emphasis on hydrology impacted the rest of my professional career, in that I focused on water resources and storm water, and not on sanitary engineering.

The issue with urbanization, then and now, is the increase in impervious area (roofs and pavements). When it rains on impervious areas the water doesn't percolate into the ground, replenishing aquifers, but rather runs off quickly into the nearest stream or river. This causes quicker and much higher flows and the resulting flooding and erosion. The weight and energy of moving water is enormous. The rule of thumb is you can stand up in water that is 3 feet deep and moving at 3 feet per second. Deeper or faster and you get swept away. Compounding the increase in the volume of runoff, is the pollution (grit, oil and grease) that washes off paved parking lots and roads. Studying the quantity and quality of urban runoff occupied much of my professional career.

My four semesters at Maryland were uneventful (September 1971 to December 1972). I worked hard, but not too hard and enjoyed their beautiful 18 hole golf course, which was empty during the week. The course was, reportedly, constructed with money from some long ago football bowl game. My best buddy was getting his PhD in chemical engineering and we played often. I have never been a good student, too many distractions.

For my work in the Anacostia River I needed to walk its entire length in Maryland, recording my observations and taking pictures. I also had to build a sediment sampler, which looked like a coffin, standing upright. I attached it to a mid-river pier on one of the bridges. It had bottles inside at different elevations and as a storm passed and the river rose, each bottle was filled with water and suspended sediment. The amount of sediment captured in each bottle was measured after a storm. With these data, calculations could be made to estimate the total sediment load for each storm.

I constructed the sampler in the work shop that served all the engineering disciplines. The work shop had every tool and piece of equipment known to man. You could have built a rocket ship there, supervised by the shop manager. He could build anything, helped me and reminded me of Phil Constantine. We became fast friends and I spent a lot of time in the shop, even when I didn't need to. We drank lots of coffee, talked about everything and played golf every so often.

I got lucky with my project in a strange sort of way. During September 1972, hurricane Agnes came up the east coast and dumped 6-7 inches of rain on the Anacostia watershed. This caused massive flooding, including washing out several bridges along the river. These bridges were upstream in the piedmont, where the river valley (flood plain) was V shaped and the water rose many feet, taking out the bridges. My bridge, where the sediment sampler was located, was downstream in the coastal plain, where the flood plain was flat and wide. Because of this the water spread out, didn't get too deep, and as a result, my bridge survived. The bridge survived, but the sampler didn't. I wasn't too disappointed, since I had most of the data I needed and now I didn't have to collect more samples. But more importantly we got to see the impact of a very large storm on sediment transport. The theory goes that most erosion occurs during large storms, and although we didn't have data from the sampler (it was destroyed in the storm), we did have pictures, before and after. They clearly showed the massive bank erosion and were exhibit A in confirming the theory.

The final report (one volume in the overall study) was well received, except I got the third degree from one of the professors (Dr. McKuen), when I orally defended the report, as the final step in getting my degree. This was Dr. McKuen's first job, with his brand new PhD. He wanted to make students suffer as he apparently did earning his degrees. With help from Dr. Ragan, I made it through. Dr. McKuen matured, wrote several books and by all accounts had a successful career.

Dr. Ragan also helped me with some sound advice when I began looking for a job. He said to work for a government agency, only if the agency had an urgent mission and funding (lots of money), otherwise it could be terribly boring. The newly formed EPA had this type of profile, but I wasn't keen on working in the public sector. In the private sector he recommended working for a large company, with lots of work options, in case the first one doesn't work out. He had received his master's degree at MIT and knew people at Metcalf & Eddy in Boston. This seemed like a good lead, since they were large, well-known and I wanted to go north to be closer to my sister, Leslie and her husband Henry, who were living in Vermont. Henry, a PhD in political science, was teaching at the University of Vermont in Burlington.

Another major reason to go north was the 10 acres of land I purchased in Duxbury Vermont right after getting out of the Army. I always wanted to build a house on the land, looking across a brook and valley to Camel's Hump, the second tallest mountain in Vermont. So, I was essentially deciding whether to stay in the DC area or go north. I wanted a city, but wasn't interested in New York City, which, although close to home, had no appeal to me. I did consider Montreal, since it was a short drive to Burlington, although not living in the US seemed a problem. The M&E lead from Dr. Ragan was the deciding factor, even though I had never been to Boston.

Like much of my life, I made the decision in my head, with little or no consultation from others, aside from Dr. Ragan's advice. More specifically, I didn't consult with my girlfriend at the time, Alayne. This was not smart, another example of my lack of communication skills, which led directly to us breaking up, as I'll describe in more detail later.

CHAPTER 3

BOSTON AND METCALF & EDDY (M&E), 1973-1985

I actually had three job offers in Boston. One with a small private start-up run by two guys from MIT, a second one from the Metropolitan District Commission (MDC), which was the public agency responsible for the Boston metropolitan area water and sewerage systems, and the third was from M&E. The decision was easy, since M&E was likely the best know environmental engineering company in the country. They had just written their latest book on wastewater collection, treatment and disposal, that was considered the bible in the industry. I still have my copy, although it's been updated and expanded several times. Metcalf & Eddy published their first edition 'American Sewerage Practice' in 1915, clearly ahead of their time, and it was still that way in the early 1970's.

M&E was started in 1909 by Leonard Metcalf and Harrison P. Eddy. Leonard Metcalf died young in his 30's, but was a renaissance man and by all accounts was an extraordinary person, beyond his engineering expertise. Harrison P. Eddy handed over the company to his son, who sold it to a larger company just before I arrived. It was a mistake, since he lost control, but it didn't much affect my life.

What did affect my life was the way M&E was run. It always had a president, but the 'Colonels' ran the place. They had careers in the Corps of Engineers, retired and entered the private sector. Two I remember were Colonels Solomon and Wilson. They were not technical gurus, but ran a tight ship, or should I say army. They knew what was going on, demanded discipline and quality control and were not to be trifled with. In short, they set the tone to achieve success through hard work.

Despite the change in ownership, my timing was perfect. M&E had just signed a large contract to evaluate the drainage system in Cleveland Ohio. It required the use of a new computer model called the EPA Storm Water Management Model (SWMM). Since it was new and sophisticated, no one at M&E, or elsewhere for that matter, had experience using it. Further, the contract had a liquidated damages clause, which stated that M&E would be fined for every day that they were late in delivering the report. As such, the principal in charge of the project, Frank Obert, was understandably very concerned about meeting the schedule. Fortunately, I had worked with the SWMM model as part of my master's project, and although I still had no clue how it worked, it looked good on my resume and I was offered a job because of this exposure (not my stellar academic performance). A perfect example of being in the right place at the right time.

We met the schedule a year later; the project was a great success and it was a huge positive start to my career. Up to this time, I was 25 years old, I had not really applied myself seriously to anything. Something in the back of my mind, maybe something my sister said, told me my "fuck-off" time was over and it was time to be a grownup. My mistakes, so far, could be forgiven and forgotten, but not anymore. This was my career we were talking about.

I certainly played a role in the project's success, but the main reason was the project engineer, Dick Reed and the Project Manager, Jek Vittands. Dick and I shared an office on the 13th floor in the Statler office building in Park Square, overlooking the Public Garden. Not bad for my first job, even though Dick, the senior guy, got the window seat. I couldn't have asked for a better boss and we worked our asses off to make the project a success. I mainly crunched the numbers, using the SWMM model, and Dick did the cost estimates and wrote most of the report. We had a few others supporting us, including the tech guys running the computers. In those days we had computer cards and sometimes there were so many cards that the model had to be run overnight. Having good computer support was essential, and we did.

Unlike today, the input data for the SWMM model was handwritten on a template in rows and columns that were later typed on punch cards by a secretary. The input data included all the physical characteristics of the drainage areas (acres, land use etc.) and pipes (size, slope etc.). There were a very large amount of data and it all had to be in the right columns on the punch cards, or there was an error. Remember, computers are loyal but stupid, at least in those days. To improve efficiency, the computer guys created a program that helped check the input data to avoid errors. It was ingenious and saved us hours trying to find simple errors, like putting the decimal point in the wrong column. I got to be good friends with the computer guys, which served me well from then on, especially when I needed a favor, or could do one for them.

Dick was an incredible manager and taught me how to be one, corrected my writing and showed me how a good engineering report was done. Lessons I carry with me to this day. He created cost tables for different types of pipe construction based on size, depth, dewatering and rock excavation. They were so good that the city of Cleveland adopted them for all their future cost estimates. They are likely good to this day, adjusting for inflation.

The key to a good report is to include enough information so the client can understand your conclusions, but not so much as to get them lost in the weeds. Keep all the gory details in the files, only to be shared if the client requests something, which rarely happened. Most projects at M&E were reviewed at a formal meeting by two or three experts who were not involved in the project. The main quality control (QC) objectives were not necessarily achieved through the review meeting, but rather by the fear of the meeting. You didn't want your head handed to you for poor work, so you prepared with extra vigor to avoid a bad review.

Two more fun facts about the Cleveland job. Dick and I had to give a progress report to the client, the Cleveland Department of Public Works. We had an early morning flight, my first "business trip". To our good fortune, we were upgraded to first class and had a really good breakfast. I was 'shitting in high cotton' thinking that this was standard fare. Sadly, it never happened again, until much, much later.

The second thing was my first 'business meeting', which was attended by the Chief Engineer and 5 or 6 staff engineers. Dick gave an overview and I spoke about how we were using the SWMM model to evaluate the drainage. Here I was, a few months out of school and considered the technical expert with respect to the hydrologic/hydraulic analysis we were doing with the SWMM model. After my presentation with lots of tables and graphs, one of the engineers asked how we were factoring in the impact of snow melt on our calculations of peak flows during the design storm. Actually, we never considered the impact of snow, so Dick looked at me as if to say 'I have no fucking idea, you are the expert, answer the question'. Not knowing the answer myself, I got up very slowly while trying to think of one. Off the top of my head I spun a great response, by noting that the snow melt would add a lot to the total runoff volume, but have little impact on the peak flow, and it was the peak flow, and not the total volume, that was used to size the new storm drains. They bought my answer, hook, line and sinker.

In private, after the meeting, Dick commented on how brilliant I was, even though I might have been wrong. The lesson, of course, was that in a public setting, you need to maintain control by being forceful in whatever you say. If you are wrong, you can always send a correction, after the fact. In this case, we looked more closely at the issues back at the hotel and didn't need to send a correction.

In addition to Dick, Jek was also an extraordinary engineer and manager. He had gone to Norwich in VT and MIT for a master's. I wound up working for him most of my time at M&E. Dick left shortly after the Cleveland job was finished. A great loss at the time, but Jek in many ways was better than Dick, if that was possible. He was far more ambitious and a good person to hook your wagon to. Jek put together a good team that stayed together for many years, including Abu Alam a PhD from MIT, maybe the smartest technical guy I ever worked with, Dan Donahue and Dave Bingham, engineers like me. Others came and went, but this was the core team. Abu was from Bangladesh and left for the US because he couldn't stand the corruption.

Just to give you an idea of Abu's capabilities, he was asked to evaluate the failure of a large wastewater force main at a big treatment plant in Middlesex NJ, a long-term client of M&E. A force main is a pipe that carries water uphill. The water is forced up the hill by a pumping station.

In this case the force main was 36 inches in diameter (that's big), 500 feet long and went uphill about 200 feet in elevation. When the pumps were running, the treated wastewater was flowing about 4-6 feet/second uphill until it reached a tank at the top of the hill. These pumps are designed to shut off slowly, but if the pump suddenly stops (e.g. power failure), there is a condition called 'water hammer'. Without the energy to force the water uphill, it slowly stops, separates, creating air pockets, and, once gravity takes over, flows back downhill toward the pump. There is normally a check valve at the pump, which closes to protect the pump. As the water comes crashing back down the pipe, it hits the check value like a 'hammer'. Water is very heavy and not compressible, so if the pipe is not designed to withstand the force of the water hammer, the pipe breaks. This is what happened in NJ.

Water hammer hydraulics are very complex, and computer programs that are used to design pumps and force mains are similarly complex, populated with many differential equations (calculus). Abu was likely the only person at M&E, and elsewhere, who truly knew the math behind water hammer models. So he was asked to do a forensic analysis of the failure. He used the most sophisticated model available to help him evaluate the situation, a model developed by several professors at McGill University in Quebec. The lead professor was an international expert in water hammer hydraulics. Abu looked at the model and found mistakes, all small, except for one that was significant. He corrected the mistakes and completed the project. He wasn't going to say anything about the changes he made to the model, so we suggested he write a nice letter to the professor, explaining the situation. He wanted to keep his improved model just for M&E, as a competitive advantage, but finally agreed to send the letter. He never got a reply, as far as I know. Many professors have great egos, unwilling to admit mistakes, especially those mistakes uncovered by outsiders.

Abu was a national treasure, but a son-of-a bitch to work for or with. He did not suffer fools and left a trail of unhappy people everywhere he went. He was my boss on several projects and I learned a lot from him, but he was the only one I ever screamed at and said "fuck you". On a personal level he was a sweetheart and, as a friend, would give you his last dollar, a characteristics of folks from the Indian sub contenant, as you will see later.

In addition to Abu, M&E has a few technical specialists, very few, but the ones they had were terrific. I'm thinking of Dave Bova (wastewater treatment) and Frank Gunby (pumps). I had the most interaction with Dave, who was a true oracle. He was approachable, asked penetrating questions, and whenever you went to him with a problem, any problem, you always came away with an answer or a way forward. He was a key reason M&E was a special place.,

M&E did water (clean) and wastewater (dirty) engineering, but because so much federal money was going into wastewater, thanks to the Federal Clean Water Act, 80-90 % of their work was in the wastewater area. Consulting engineering has two basic components: planning and design. The bread and butter work for M&E was planning and designing wastewater treatment plants. Since, as mentioned above, they wrote the standard textbook on the subject, M&E was the place to be in the 1970s, when I arrived.

The Clean Water Act focused on improving wastewater treatment, but also had a number of programs designed to improve the nations waters in a comprehensive way. These included river basin planning (think the Ohio River), areawide wastewater planning (thing the Baltimore metropolitan area) and combined sewer/storm water planning (think Boston when it rains). The idea was to look at the big picture and make sure the work at a particular treatment plant fit with other state and regional needs. For example, in Boston, if you just fixed the treatment plants and not the pollution from combined sewer overflows (CSOs), the harbor would still be polluted when it rained, and many days after. So you needed a plan to address both dry and wet weather pollution.

Our group, run by Jek, focused on river and areawide planning and wet weather pollution (CSOs, storm water). If we had a project that went from planning to design, we typically handed it off to other engineers, who loved the detailed work of design. We were big picture guys. So we wrote a lot of reports. The good news: it's hard to get sued for negligence in a report, because mistakes can be fixed in the final design phase, before construction.

There were many reports, but perhaps the most consequential was a master plan we did in 1978-79, for the Metropolitan Boston Sewerage System. It was called the EMMA Study, for Eastern Massachusetts Metropolitan Area, which covered about 40 towns, including Boston. This was before Mike Dukakis ran for president and lost, partly because George HW Bush pointed out that Boston Harbor was the most polluted harbor in the country. He was right, but it wasn't Dukakis' fault. Our report was ten volumes and proposed a plan to clean up the harbor. Too bad for Dukakis it wasn't implemented before he ran for president ten years later. The problem, which was/is common in Boston, was consensus and cost. There wasn't agreement on the right approach and every solution was very expensive. Finally, in the 1990s, the courts stepped in and ordered a solution, but let's get back to my role.

One of the senior VPs in the company, who will not be named, was in charge of the EMMA Study and had been running it for about nine months. The problem was that most of the fee had been spent, but little useful work had been done. Since this was a very high profile project for M&E, there was panic in the C suite. The solution, direct from the president, was that Jek and his merry band would take over the project and fix it. To the president's credit, he wanted a quality product to protect our reputation, knowing full well that the company would take a significant financial loss. This was clearly the right decision from both a professional and business standpoint. A lesson that I never forget, although the main lesson is to avoid getting in a bad situation in the first place.

Jek got us together and said we needed to work around the clock for the next 4 to 6 months and produce a world class report. The company was depending on us and, although unsaid by Jake, if he pulled it off it would be a ticket for him to the top. We actually got excited and delivered, although some of the volumes were delivered a month or two late and many overtime hours were never charged to the job to minimize the financial loss.

I was responsible for the Combined Sewer Overflow (CSO) volume, one of ten volumes, since I had done similar projects in other cities. All the team members mentioned above were involved in other volumes. In addition, we had help from Ara Chamlien, Steve Koop and Gene Davis, who I got to know well. We really had a good team and great moral. The client was thrilled and very tolerant about the delays, knowing that the quality was excellent and the proposed solutions were what they wanted.

Since I spent so much of my time on studying combined sewers, a brief history of CSO control in Boston is warranted. First the problem. Today, if you were planning a city you would build sewers, to carry sewage, and drains to carry storm water runoff. Two separate pipes. However, 100 years ago, when many older northern cities from Chicago to Philadelphia to Boston were building this infrastructure, they chose to build one pipe to carry both sewage and drainage. These pipes needed to be large to carry runoff from heavy rains, while the everyday sewage flows were much smaller and only needed a small portion of the pipe's capacity.

Originally, there were no treatment plants and the combined flows (sewage every day and storm water infrequently) were discharged to the nearest river, lake, or harbor. When treatment plants were built, starting in the 1940s and 1950s, all the discharge points were connected to an "interceptor sewer" which carried dry weather flow to the plant. The treatment plants were designed to only treat the sewage, so the interceptor sewers were relatively small compared to the combined sewers. Consequently, when it rained the interceptors weren't large enough to capture the combined flows, so there were overflows at each connection between the combined sewer and the interceptors. These overflows occurred 50 to 100 times a year and contained sewage and runoff, polluting the receiving waters.

This is what happened in Boston. By the 1960s there were two treatment plants discharging to Boston's inner harbor, every day 24/7, and 130 combined sewer overflows discharging to the Charles and Neponset Rivers and the harbor when it rained. In the 1990s, the southern plant on Nut Island was eliminated and connected to the northern plant on Deer Island which was expanded and upgraded with more advanced treatment. The treated wastewater was then discharged through a 9 mile outfall 100 feet deep in Massachusetts Bay. This solved the dry weather problem. The wet weather CSO problem was more complicated and was subjected to five iterations of planning, starting in the 1960s.

Because CSOs are infrequent (when it rains) and the flows change dramatically over time, it is difficult (very high flows over short periods of time) and expensive to treat the overflows in real time. Consequently, there are two basic approaches to control CSOs. The first is to provide storage to capture the overflows, and after the rain stops, slowly, over several days, bleed the captured water back into the interceptor system to be treated at the treatment plant. The second approach is to separate the combined sewers. This usually means building a new sewer system and leaving the combined sewer to collect only storm water. These two approaches have been argued over for the past 60 years.

I always favored storage over separation, because I thought, correctly, that even if you separated the combined sewers, the regulators would eventually make you treat the separate storm water. I had my battles with John Sullivan, Chief Engineer of the Boston Water and Sewer Commission, over this issue. He liked separation in part because when the streets were dug up for new sewers, he could fix other utility issues (e.g. water mains) and give the neighbors a newly paved street.

At first the storage approach won and then times changed, and sewer separation became fashionable, as the list below of proposed solutions illustrates:

1. A Deep tunnel storage plan to capture CSOs was proposed in the 1960s.
2. A series of surface storage tanks to capture CSOs was proposed in the 1970s as part of the EMMA study that I worked on.
3. Another deep tunnel plan was proposed in the early 1980s.
4. Another surface storage plan was proposed in the late 1980s, which I also worked on.
5. By 2000, there was a shift toward sewer separation, and much separation was done in Boston to address the problem.

6. In 2005, a tunnel was constructed in South Boston to capture the overflows and protect the M and L Street beaches off of Day Blvd.

Through all my time at M&E we were the go-to group for CSO work, and later storm water work. This became my technical specialty, and I was lucky because it was a new field with not much history. I published several technical papers on CSOs and storm water, and got to go to a number of wonderful conferences all over the country. Technical folks used the terms, wet weather, storm water management or non-point source pollution as the technology developed.

In the early days, before my time, the engineering goal was to get the storm water off a site as quickly as possible. You designed catch-basins and storm drains directed to the nearest river, stream or ditch to prevent on-site flooding. This was fine for the site, but began to cause downstream flooding. Next came the need to control storm water on the site so as not to worsen downstream flooding. This was typically done with detention basins that stored runoff and released it slowly. All this dealt with the quantity of storm water runoff. The last stage was the realization that the quality of storm water runoff was also a problem. This became clear as point sources of pollution (e.g. municipal treatment plants) were improved and the receiving waters were still polluted by storm water. The problem was amplified with CSOs, because they carried both storm water pollutants and sewage.

When I came on the scene, there was not much technical literature and design criteria for storm water pollution control, so we made it up as we went. I think our biggest contribution was setting the design criteria for controlling CSOs. Without getting too technical, we said that CSO pollution control facilities should be designed for overflows from a 1 year, 6-hour storm (the worst rainfall that occurs every year for 6 hours). In Boston that was about 2 inches but it's more now with climate change. Because CSO control was so expensive, we based the design criteria on a crude "knee-of-the-curve" economic analysis, which illustrated that you got the biggest bang for your buck from this design.

Or said another way, the added cost for designing for a bigger storm was not worth the pollution gain. There was a lot of judgement here, but the logic held together and no one else had a better idea. The hard-nosed environmentalists wanted zero pollution, but that was economically out of reach for CSOs. Over the years, everyone seemed to adopt our design criteria, including EPA, not knowing where it originated. We developed and used it first for a CSO project in Middletown, Ohio, and it stuck.

Speaking of Ohio, we had a world class job of modeling the entire Ohio River from its start in Pittsburg to the Mississippi River, hundreds of miles. This was funded by the Clean Water Act. The goal was to determine what level of treatment was needed at each major municipal and industrial discharge to maintain acceptable water quality in the river. We hired George Schroeder, a professor at the University of Minnesota, to help. He was a wise old man and submitted his reports in long hand. He was brilliant, once you figured out his penmanship. He helped us select the inputs to the hydraulic and water quality model we were using. Great project.

One of our main client contacts at the Boston Water and Sewer Commission (BWSC) and later the Massachusetts Water Resources Authority (MWRA) was Libby Blank. She was a planner and one of the few well-known women in the field. She was known to be a difficult client, but I got along well with her. I think this was partly because when we went to public meetings I could talk to people about technical issues in language they could understand. I also did well with questions and answers. In short, I represented our client (Libby) well and made her look good. This was likely my main strength as a consultant, not only with Libby, but with others.

Perhaps JFK described the lot of a consultant best, when he said to one of his advisors " If we are successful I take the credit, if we fail you fall on your sword". I have found that most consultants don't mind falling on their sword, when a mistake is made, but go crazy when others take credit for their good work. You need to get use to this, because a happy client will give you more work, and that is what it's all about.

I grew to love the work and got good at it. I traveled to clients well beyond the Boston area, including Cleveland and Middletown, Ohio, Birmingham, AL, Savannah, GA, Tampa, FL, Frankfort and Lexington, KY, Orange County, CA, Middlesex and Cape May, NJ, Wilmington, NC, Granby, Quebec, and a lot more work in Boston.

I was single, so I often worked well into the night and on weekends. A habit that continued for the rest of my career. Part of this drive was my feeling that I was the dumbest guy in the room, so I had to work harder to keep up. Remember, M&E was a major league team. Whether right or wrong, this was a good approach, one I never regretted.

There was/is no way to succeed in the consultant engineering world without a professional engineering (PE) license. To get licensed, you need to pass two 8-hour exams. The first is the engineering in training (EIT) exam, which you can take anytime. It's recommended you take it at the end of your senior year, because it covers all types of engineering problems, not just civil engineering. Of course, I didn't take it in college because it was given on a Saturday in May when the weather was nice, and I had other interests. Stupid, very stupid! You needed a PE to survive at M&E, or anywhere else. When I realized this, I took the EIT on a Saturday during my first year at M&E. At lunch I seriously considered not going back for the afternoon session, given what I thought was my poor performance all morning. Fortunately, I went back, apparently did better in the afternoon, and passed.

The next step was the PE exam, but you needed 4-5 years of experience before being eligible. You were licensed by state, and each had different requirements. You could take the exam in Maine a year earlier then Massachusetts, so everyone did, after being approved based on an extensive application with references. If you passed the exam in Maine, you could ask for reciprocity in Massachusetts by completing an application. The major benefit was you didn't have to retake the exam, which was very similar state to state. Most of us took this approach.

Of course, the exam was given in Maine, so we used an M&E car and they paid for our hotel room the Friday night before the exam in Augusta. I was driving with three other guys on the Maine Turnpike when I got a speeding ticket, which M&E did not pay. Not a good sign for the exam ahead.

The EIT exam had a few hundred multiple choice questions, while the PE exam required you to answer 8 questions you picked from about 50 choices. The answers were not multiple choice; you submitted all your work and calculations. This exam was focused on civil engineering, but that includes other disciplines besides water/wastewater, such a structural bridge design. In any event, I was smarter this time, took a refresher course, and got a passing grade, which was all you needed.

I got the results several months later. I had come home for lunch and saw the mailman walking up the street while sitting at the kitchen table on Mt. Vernon Street. Some of the guys had gotten results so I knew mine was coming. I ran down to the front door, opened the letter, and was as happy as I had ever been. More so, because Jek had pushed through my early promotion to project manager, and told me he would be embarrassed if, after that, I didn't pass the PE exam. Your PE license was your ticket to get on the train, or, in my case, to stay on the train. I can't overemphasize the importance of a PE in the consulting world, then and now!

To be in the consultant business you clearly needed your own credentials (e.g. college degree, grad school, PE), but after those threshold requirements you needed to be able to work in groups. Consultants do projects, and projects are done by groups. You need to understand group dynamics (people) to be successful and when you get to the point of managing projects, you need to be able to attract good people who want to work for you and get stuff done as a team. In short, you need some level of emotional intelligence to work in groups. Technical ability is not enough.

In addition to the technical folks, you need to understand how other departments can help you. Financial folks will help develop cost estimates and collect money, marketing folks will help with proposals and interviews, IT folks will help with computer programs and communications, HR will help with hiring and personnel issues. You need to work with all these people because they can make the difference between routine and excellent results.

The project manager (PM) is the key person, the heart, of a consulting business. She gets work, executes the work (with a team), and collects the money. You spend your first 5 to 10 years preparing to become a PM.

During this time, you get your PE license and build your reputation, typically by focusing on a technical area and respecting people. You also join professional organizations, write technical papers, and get to know your clients. The goal of most young engineers is to become a PM, where you shift from doing calculations to managing those who do them.

The biggest change when you become a PM is the responsibility for bringing in new business. This is the life blood of the company. The PMs who are good at it are the ones who become vice presidents and leaders of the company. These folks are called rainmakers and are prized in the industry, even though they fundamentally use the same skills they learned as a PM to get things done. In the end you need to get the technical stuff right, but the difference between the good and great companies are the connections you make with people, especially the people who become your clients. The best way to keep your clients, and get new ones, is by doing each project well, on time and on budget.

One last point about connecting with clients. No one is perfect and you are bound to make mistakes, sometimes big ones, but it's not the mistakes that bring you down, it's how you address them. Taking ownership of mistakes and being transparent and fixing them in a timely fashion will strengthen your relationship with clients. If not, you should be looking for different clients.

3A. THE GOOD LIFE

There was more to my life than work. I had a fourth floor walk-up apartment at 175 Beacon Street, a 10-minute walk to the office. I became a city boy. Dad would have approved. The first person I met at work was the HR guy, Paul Wold. He became my best friend at M&E and gave me very good advice that first day about getting an apartment in the city, rather than the suburbs.

Living in the city allowed me to sell my car, since I could walk back and forth to work and also to the local bars. Paul introduced me to his friends, not from M&E, and we spent our time mostly at the bar in the basement of the Hampshire House, now called Cheers, a short walk across the Public Garden from work. Alayne came to visit and apparently didn't approve of my life style and drinking buddies. Being clueless, and in my own world, I assumed she would eventually follow me to Boston, despite my wayward ways. Well, I was wrong, even though I wasn't emotionally or financially ready to get married. When her school year was over, she told me she had accepted a job at a Lutheran school in St Louis. I was shocked and walked around in a bit of a fog for a few days. After all, we had been going together on and off since high school, and this certainly seemed like the end. Paul took me to the Hampshire House to drown my sorrows, and I survived.

I was new to Boston and without a car, so I didn't get out of downtown often. That was partly solved when M&E launched a softball team. We were called the Rotifers, after a bug that eats sewage in a wastewater treatment plant. It wouldn't have been my first choice, but it stuck. Fortunately, no one else knew what it meant. We played on the Common, my favorite venue because of the short right field fence, but also in many other neighborhoods as part of a City of Boston league. I would hitch a ride from Paul, and we would explore the neighborhoods on the way to and from the games: South Boston, Dorchester, JP, Roxbury, West Roxbury, and others. It's how I got to know the city and several of my co-workers. Russ Adams and I were the best players and, as left-handed hitters, enjoyed the fields with short right field fences. Some of the locations were not so hot. One night playing near Northeastern, we heard gun shots coming from a nearby building, followed by police sirens.

During this early time in Boston, I was living in a studio apartment on Beacon Street and Peter Swartz was my landlord. He was a few years older and had gone to Colby College and Harvard Business School. Although he was very smart and unconventional, unlike me, we became friends. He owned several buildings and soon we were looking for one to buy together and live in. To make a long story short, I finally bought a building by myself, since it was rent controlled and Peter didn't buy rent-controlled buildings. It was a steal, since the owner was filing bankruptcy. I only needed $3,000 in cash to pay his debts and assume the mortgages. I borrowed the 3K from my mother, which was the good news. The bad news were the 2 mortgages, one with a bank and one with a private lender. The bank mortgage was typical, but the private one was at a 16% interest rate. Needless to say, I rolled over both mortgages as soon as the terms allowed.

Even given the mortgage situation, this was the best deal I have ever made, by far, especially since the building was a premier Beacon Hill property on the best street, Mt. Vernon. Built in 1831 as a single-family home for Phineas Upham, with large rooms, high ceilings, and extensive woodwork. It had a bay front with a marble fireplace in every room and was set back from the street with a beautiful front yard. A year after I bought the building, the city eliminated rent control.

One day an old friend of the family stopped by and we were sitting in my living room, which was then on the first floor. This was an impressive, large bow-front room with tall windows, a beautiful marble fireplace, high ceilings, chair rails, and inlaid hardwood floors. The door to the foyer was open and we were chatting when one of the upper floor tenants walked in to pay her rent. She happened to be this gorgeous blond, who smiled and handed me the check, as my friend's eyes were popping out. After she left, he caught his breath and said he wanted to move in. Sadly, for him, he was married with two kids and lived in upstate NY.

When I bought the building it was a rooming house, with three rooms and a common bath on each floor. There were two kitchens, one in the basement, which I used, and one on the fifth floor. I converted each floor into an apartment, by adding kitchens. This happened over two years when I worked on the changes almost every weekend, with help from plumbers and electricians. I did a lot of painting.

After painting all day, I had some good nights out. One time I was at the Hampshire House a bit too long with a lovely lady, Robin, who was my neighbor, living in Peter's building at 73 Mt. Vernon. She had come to Boston from Albany, where she had an affair with her boss, a state senator. She was a hot ticket and we had some good times. This night, actually early morning, we were "three sheets to the wind" and walked up the middle of Mt. Vernon Street singing songs and quoting Kipling's "Gunga Din" and "The Road to Mandalay". No cell phones to capture quite a scene.

I tucked Robin in bed and stumbled home, two doors away. I walked through the two front doors and the foyer door to my bed on the first floor, neglecting to shut the doors behind me. I passed out on the bed and then during the night two unpleasant things happened, without me knowing. In no particular order, I threw up and someone walked right through the open front doors and stole my TV, camera, radio, and record player. Fortunately, I didn't have too many valuable worldly possessions at the time.

I woke up the next morning to banging on the front door (and in my head), and the calling of my name. Turns out that Fran Bendler, my mother's good buddy, was in Boston with a friend for a Christian Science meeting. My mother had mentioned this, but of course, I had forgotten. Fran and her friend were stopping by to say hello. I quickly changed my shirt and asked them in, trying to sit as far away as possible and down wind. Fran was a really cool lady, had two sons like me, and knew exactly what state I was in, as I explained it had been a long night. During our conversation I started to get my wits about me and began to notice that the TV etc. were missing. It became clear when Fran pointed out that all the doors were wide open. Robin came over to comfort me, but with all my shit stolen, it was quiet around the apartment for some time.

Robin and I didn't last long but remained friends. I took her to one of the M&E Christmas parties and she flirted with several senior folks, much to my advantage. I had a string of very good-looking dates to M&E Christmas parties, two others being Eastern Airline stewardesses. From what I heard, this impressed a few VPs, and certainly didn't hurt my corporate advancement. You need to use all the tools in the toolbox.

I was friends with Wes Books, a neighbor when I lived at 175 Beacon Street. He had a share of a farmhouse, with two other guys, in Rochester, Vermont. We went up a number of times and I would often make a side trip to the property in Duxbury and check it out. We always tried to convince two lovely ladies to make the trip with us and were often successful. We would fish, shoot skeet, canoe, and generally had a good time. In the winter the girls would ski, and we would wait at the bar. Good times. Wes was a good cook and would make culinary masterpieces in the large eat-in kitchen. The only trouble, dinner was never ready until 9 or 10 PM, and by that time everyone was too drunk to enjoy the food. One summer Pete Brubaker and I went to the Duxbury land and camped out. Although I didn't see Pete much in those days, we became close friends and later had many adventures together.

I few years later I met Cynthia at, of all places, the Hampshire House. She quickly moved into 69 Mt. Vernon and a year later we got married. Keep in mind that the building was a rooming house when I bought it. At first my kitchen was in the basement, and I lived on the first floor. This all slowly changed by putting in apartments, with ours on the second and third floors. It was truly, high end city living, at one of the best addresses in the city.

Again, fast forward a few years and Erin was born, and we bought a summer cabin on Thomas Pond in Raymond, Maine. Even though I had land in Duxbury, Vermont, the ride to Maine was much shorter, and, once there, you walked a few feet out the cabin door, sat on the dock, and looked across the lake while fishing. Cynthia was not a big traveler and loved Maine as I did, so we spent summer weekends and our vacations all in Maine. Good decision, although the Vermont land was never developed.

After Alayne left for St. Louis and before I met Cynthia, I was single and lived a carefree social life, as noted above. I hung around mostly with Paul, Ara, and Steve. Paul was my drinking buddy, Ara was my backgammon buddy, and Steve was my squash buddy. That's an oversimplification. I did a lot of social stuff with Paul, often with his beautiful future bride from North Carolina. With Ara, it was often dinner in Cambridge at Casablanca. He also helped me move from Beacon Street (4th floor walk-up) to Mt. Vernon Street, for which I will forever be grateful. Steve was a joiner, so we joined the Tennis and Racket Club on Boylston Street. It had several squash courts, a racket court, and a tennis court (the original Henry VIII version and one of only a few in the country). We played a lot and on one Thursday night after playing, Steve dropped me off at the Hampshire House on his way to the train. It was the night I met Cynthia.

Once I got married and Erin was born, things became more predictable and stable. Cynthia worked at New England Nuclear and valued her job, so when it was time to have a family, there was much discussion over how things would work. I was more interested in starting a family, so I suggested that I stay home for a year, or two, when the baby was born and Cynthia would continue to work. She thought that was a great idea, despite the loss of my income, so that's what happened.

At work, when I told Jek what my plan was, he was very supportive, actually saying that he should have done it himself when his daughter was born. The HR people (Paul had left by this time) weren't so enthusiastic. They pointed out that there was no guarantee that I would have a job when I wanted to come back, and further, they wanted no written record of my reason for leaving. They were afraid that the idea would catch on, and other engineers would leave. Staffing was a constant industry problem in a booming environmental market, and almost all the engineers were men. When one woman engineer started, a senior VP asked her to type a letter, assuming all women were administrative staff (secretaries). In 1980, M&E was not a very progressive company, but I had no problem with their conditions upon my leaving. It was one of my better decisions. As it turned out, M&E had nothing to worry about, at least in the short term. Few other men took my lead in those days. I was a bit ahead of the times.

I cherish the time, about 18 months, that I took off to stay with Erin. She was too young to notice much, but I loved walking all over the city with her in the stroller. In the good weather we spent time in the Public Garden and Maine. My friend Peter's wife, Marsha, had two babies and lived two doors away on Mt. Vernon Street. One day I suggested we go together to the playground along the Charles River, with the kids. She nicely pointed out that the other women would feel threatened with me there, in part because if their husbands found out they would be pissed. Also, a million rumors would start flying around about who I was trying to get in bed with, while the kids were napping. Boy, was I naive.

3B. MORE WORK STORIES.

After my time at home, I was ready to go back to work, so we sent Erin to daycare at the Park Street church. I could now appreciate how a woman could feel caring for young children at home for many years. M&E took me back and I was up and running in a few months, as if I never left. I didn't forget how to do the work, just like you don't forget how to ride a bike. I missed two pay raise cycles, but that was part of the deal.

When I started back, the biggest change was my relationship with Jek. He had gotten a new secretary, divorced his wife, and married the secretary. She worked at a desk outside his corner office, and you had to go through her to see him. She made clear who controlled access to Jek. Certainly, not like the informal old days. This was very distasteful for me and sadly suggested that Jek was on his quest to become the president of M&E, which he eventually did. I instinctively looked for other senior people to associate with.

For the first year back, I worked on two similar jobs in New Jersey. They were both part of a federal program (301h waiver) that allowed municipalities to apply for a reduced level of wastewater treatment, from standard minimum requirements (e.g., secondary treatment), if they discharged to the ocean. We were contracted by cities from NY Harbor to Cape May to provide the technical reports to support their requests for a reduced level of treatment, and, of course, save a lot of money in the process.

This program worked well on the west coast, where the ocean becomes very deep just offshore and there are strong currents to dilute the wastewater, after a modest amount of treatment. I worked on the one for Orange County, CA, just south of LA, and it was eventually successful. On the east coast, where there is shallow water on the continental shelf, for miles offshore, the conditions are not favorable.

Nevertheless, there was so much money to save that the east coast cities, including Boston, all wanted to apply. So, as good consultants, we took their money and produced huge reports making their best case. We knew in the end that their application for a waiver would likely be denied. All were, including Boston's. In South Jersey, it was a no brainer, given that their economy ran largely on beach tourism. Think Atlantic City, before the casinos.

Despite tilting at windmills, the work was fascinating. It combined oceanography (current, temperature, salinity), water chemistry (organic decay, dissolved oxygen), biology (fish and benthic organisms) and public health (bacteria, viruses). It was everything an environmental engineer could want, including field work measuring water chemistry and biology. Pulling all the information together was a tour-de-force in project management and I did two of them. Clearly, I was back in the game.

Like most of the work at M&E, and elsewhere, producing these reports was a team effort and, as a result, I got to know two wonderful, very competent people, both PhDs, one scientist and one engineer. Bob Reimold was the scientist, who got his PhD in biology at the University of Georgia, under famed ecologist Eugene Odum. Bob was a nerdy scientist, compulsive about details, but with a great sense of humor. He always delivered excellent work, on time and within budget. You had to love the guy.

As I mentioned there was field work involved, including one expedition off Cape May, NJ to collect water, fish, and bottom sediment samples. Bob loved field work and arranged the whole affair, boat included. I didn't need to go, but as project manager, I wanted to show I was deeply involved in every aspect of the project. Plus, it was a summer day on the water.

There were to be three days collecting samples, but I only went the first day. That was enough to show my interest. Bob had rented a good-sized fishing boat, with a big winch on the back to lower heavy equipment into the water. There were five of us to do the work, but only four were needed. I was the boss, but, in this case, only provided an extra set of willing hands. We left at dawn and worked hard all morning, sampling the water and bottom sediments non-stop. In one process, a small dredge picked up a bottom sample of sand and mud and dumped it on the back deck of the boat. Our biologists, including Bob, sorted out the mud and identified the benthic critters (worms, shrimp, and more). Because the bottom critters can't move very far, they can't avoid pollution by swimming away, and, as such, provide a good indication of the health of the local waters. In this case they showed a healthy environment off the southern coast of New Jersey.

Inside the boat there was a kitchen and eating area. At lunch the cook set out a feast, since, by that time, we were all starving. I ate and drank like there was no tomorrow, which turned out to be a mistake. Although it was a perfectly beautiful, calm, sunny day, I started to feel seasick. To avoid embarrassment, I walked to the front of the boat, laid down, and tried to take a nap to settle my stomach, out of sight of the rest of the crew. It worked, but while I was spread eagle and napping on the front of the boat, Bob took my picture. Like a good Boy Scout, he was always ready for the moment, with his Swiss Army knife, binoculars, and camera.

Bob didn't mention the picture to me, but a few days after they got back to the office, the picture appeared on the company bulletin board. I took a lot of grief about sleeping on the job, but it was all in good fun. A number of years later, after both Bob and I had left M&E, he asked if I would contribute a chapter in a book he was editing on watershed management. I said yes, immediately, knowing if Bob was involved, it would be of high quality. I also had a good project to use as a case study: managing the Hobbs Brook watershed, which is the drinking water supply for the City of Cambridge, MA. The book was published in 1998, entitled *Watershed Management, Practices, Policies and Coordination,* edited by Robert J. Reimold, PhD. I was right about Bob's attention to detail and his high standards. Sadly, a few years later Bob died suddenly, in his 50s, way too young.

As if Bob wasn't good enough, I also had a new M&E engineer assigned to me. The engineer was Dave Langseth, with a PhD from MIT. He was right out of school and didn't have much consulting experience but was one of the three smartest engineers I have ever worked with. The first one was Abu, who I have already mentioned, and the third one will show up later.

We had to run hydraulic and water quality computer programs as part of the 301 waiver applications and Dave did all that complicated work, flawlessly. Like me, at first, he needed work in polishing his writing, but with some of my editing help, he picked things up quickly. He not only ran the computer programs, but like Abu, knew how the equations worked and could critically analyze the results.

In another project, for the state of Vermont, we needed to use a computer program that had been developed by EPA. Dave went "under the hood" to see exactly how the program was developed, and made several changes and corrected several bugs. These proved invaluable in completing the project successfully. Unlike Abu, Dave had an excellent bedside manner. Although he was smarter than others in the room, he didn't talk down to the people he worked with. In the case of the EPA program, he called the developer, who was a nationally recognized expert in the field, and calmly explained our project and what he had done to the guy's program. The expert, who I knew from other work, was a good guy and delighted to get the information. He went back to EPA and made the changes, so others would benefit. We actually put his name on the final report as an outside consultant, and everyone was happy.

Dave, like most of us, left M&E for somewhere else. In his case it was Northeastern University, where he was a civil engineering professor for many years. Every year, seniors had a capstone project, which was intended to be a real-life engineering problem. I helped him with the capstone project several years and enjoyed my interaction with him and the students.

In the end, I was blessed to have worked with both Bob and Dave.

As noted above, a large part of my work was related to running computer models. It started by running rainfall/runoff models (e.g., SWMM) to design storm drains and combined sewers. It moved to receiving water models to predict the pollution impact in fresh and marine waters. Some included the Charles River, Ohio River, Great Miami River, Savannah River, Hillsboro River, Connecticut River, Boston Harbor, Raritan Bay, Atlantic Ocean off South Jersey, Cape Fear estuary, NC, and Tampa Bay, FL.

Receiving water models generally have two parts, hydraulic and water quality. The first, hydraulics, is the easy part. Water is incompressible and follows well-known physical principles. Water in a river flows downhill based on the channel configuration, and marine waters move according to the tides. Water movement is predictable and easily verified with flow gauges and tide charts. Things get a lot more complicated when trying to predict pollution impacts based on chemical and biological reactions. There are many, many variables that influence a natural environment, including temperature, pH, dissolved oxygen (DO), conductivity, bug metabolism, and many more.

When doing a receiving water modeling project, you can usually get a robust hydraulic model that you have confidence in, but not so much the water quality portion of the model. That is why it is so important to report your results not only as a single, discrete answer, but how confident you are in the result(s). For example, if the goal/standard is to keep at least 5 mg/l of DO in the water at all times, and your model predicts 4.5 mg/l under design conditions, is that a problem? To answer the question, you need to know how confident you are in the 4.5 mg/l number. If you are 90% confident you are between 4 and 5 mg/l, you likely have a problem. If you are 90% confident you are between 3 and 6 mg/l, you may not have as much of a problem. The point is, decisionmakers need to know the level of confidence in the model results when they are asked to commit millions of dollars to correct a water pollution problem that may or may not exist.

As things progressed after I came back to work, I was getting away from my association with Jek. On the NJ waiver applications, I was actually working for the VP who headed the M&E office in NJ. He was a good guy, but had little involvement as long as things were going well. Fortunately, they were. My last big job at M&E also didn't involve Jek. I can't remember how I got involved, but we were a subcontractor to SRI International in Palo Alto, CA. They were the old Stanford Research Institute, which was independent and no longer associated with Stanford University. The project was for the National Weather Service (NWS) as they were deploying the next generation of weather radar (NEXRAD).

There was a competition between Raytheon and Lockheed to see who would design the radar. It was Doppler radar, which was new and more sophisticated in detecting wind patterns. They were to be located at airports to help detect wind shear and help planes avoid these strong down drafts, which had recently caused a commercial crash, with over 100 dead, at Kennedy Airport in NYC. They were also to be the main radar for general weather forecasting. As such, the plan was to have about 60 of the radars located at airports all across the country, including Alaska and Hawaii. The SRI/M&E mission was to evaluate each site for the suitability of the radar (SRI) and the infrastructure to support it (M&E).

I got involved in preparing the proposal, and surprise-surprise we won. It had nothing to do with water pollution, but I inherited the project manager (PM) role for M&E. At first, I was looking for a way to back out, but I got to know the SRI project manager and we hit it off. He was from Brooklyn. There were also benefits. The NWS Manager liked to travel to SRI for progress meetings. I also loved going to San Francisco/ Palo Alto, but for different reasons. He was a bit of a space shot and loved to meditate in pods filled with liquid. The liquid was dense, so you floated on top, partly submerged. You climbed in naked, the lid was closed, and you floated in the dark, while meditating. He asked me to go with him once, but I declined, not liking dark, confined places. Palo Alto was one of only two places in the country with these pods. Only in crazy California.

There were other benefits. The budget was very large and I essentially had the ability to fly anywhere in the US to keep track of the project. We needed to do site assessments on all 60 airports to make sure each had vehicle and utility access. If not, we needed to find another suitable site for the radar on the airport. I set up the system and did the first site assessment. Once the system was set up and the bugs worked out, I sat back and had three young project engineers do all the other sites.

The best part was the budget allowed for an administrative assistant/ secretary. I hired a young, beautiful, blonde woman from Chelsea, who wore clothes much to her advantage. She was a sweet, hardworking kid, completely loyal to her boss, me. The funny part was that she had a working-class accent that didn't go with the way she looked. I almost didn't hire her because of it but couldn't resist. At M&E at the time, senior VPs had secretaries and VPs generally shared a secretary. So, here I was, a lowly PM with a secretary, and she happened to be the most beautiful woman in the building. This was not received well in all parts of M&E, but I didn't care. We became friends and laughed at the reactions of others.

The downside was, after the start-up, things were boring and I wasn't learning anything in the technical area I was interested in, namely dry and wet weather water pollution. Also, at the time M&E decided to move out of Boston to Wakefield. I had to look at a map to see where Wakefield was located. So, I left the day M&E moved to Wakefield, over Thanksgiving weekend in 1985. Jek offered me a significant raise to stay, but I had made up my mind and I don't believe it's good form to change your mind for money, at least not the amount of money they were talking about.

3C. THE GUYS.

I was blessed to start my career at M&E, especially since the federal government had just opened the coffers to fund the pollution control work that firms like M&E were doing. M&E was one of the top 2 or 3 environmental engineering companies in the country, and it was the place to be, mostly because of the people. I have mentioned some of the best, who taught me 90% of what I know about engineering and the consulting business. The people were amazing: technically, professionally and personally, and for the most part, easy to work with.

Unfortunately, over the years, so many people left, as M&E was repeatedly sold and re-sold and finally, dissolved into a bigger firm. Once the venture capital folks get control, typically by buying a company with borrowed money, things change. This is what happened to M&E after I left, where a professional culture changed into a bottom-line culture. This isn't always bad, but in the M&E case, it was the beginning of the end. I believe the time I was there, in the 1970s and 1980s, was their golden age, for many reasons, but not because of me. I have remained friends with many of my old colleagues, quality people. Four of us still play golf, so let me use them as examples of why we all contributed to the golden age.

Robb Googins, a physics major from the University of Maine, did some project work, but got his MBA and became M&E's CFO in the 1980s. He left in part because of concerns about how the VC folks were managing the financials and went on to be CFO/CEO for several other companies, all consulting businesses. His timing always seemed excellent. Each job Robb had seemed to be a step up, in the right direction.

Robb is a renaissance man, who introduced me to the Philosophy Foundation, where he was very active for decades. Their teachings are based on Eastern philosophy, especially Hinduism and early Sanskrit texts. I was intrigued with their approach to life and took several courses but couldn't commit to the involvement they expected. Partly because of the influence of the Philosophy Foundation, Robb meditates every day, but not in the California water-filled pods. I credit his meditation, in small part, as the reason he always provides wise advice in a quiet, non-judgmental way. He is a gentleman who looks at the big picture and what is right. You can trust Robb with your firstborn. He also has more holes-in-one (3) than any golfer I know.

Mike Hanlon, a civil engineer from Northeastern University, who, like me, did a lot of different things at M&E. His first job out of college was for a transportation firm, then he went to M&E and stayed in environmental engineering for the rest of his career. Mike took a break from New England in mid-career and followed his girlfriend to sunny Florida, where he went to the dark side and worked in the public sector for the city of Orlando. Finally, he went back to private consulting, before coming back to Massachusetts. When he came back, he worked for an old-line New England firm, Weston & Sampson, where he spent the rest of his career, soon running the place, retiring as their CEO.

Mike is a no-nonsense type of guy, who likes to make decisions and get things done. When we take a trip, be it canoeing on the Allagash River, biking along the Erie Canal, or golfing in Florida, Mike has a plan, is organized, and well prepared. He is self-reliant and it's a pleasure to travel with him. Because of this, I have always wondered why he gets so involved in town and condo politics, where the frustration quotient is high. Opinions are many and things take forever to get done. We often hear his frustration, but he cares, and that's why he is involved. Mike is one of the good guys, whom you can count on to stand by your side and tell the truth. My only complaint is that he takes too many mulligans on the golf course.

Stan Elkerton, another civil engineer, worked in M&E's sweet spot, designing wastewater treatment plants. He went to Clarkson University and was a rising star at M&E but left to run his own operation. In his case, the Boston branch office of a larger firm, headquartered in North Carolina. Later, he and a partner started their own firm, and, eventually, near retirement, sold it to a big national firm. He was an engineer's engineer, and later did expert testimony. The lawyers loved his quiet competence. Stan has also served on the Board of professional organizations, local municipal zoning boards, and non-profits. He is generous with his time.

Unlike many technical people, Stan is a people person. He can walk into a bar and talk to the wall but would much prefer to talk to an attractive woman, and does. His standard opening line is, "where did you go to school?" It works, since he is a wonderful person and a good listener, who is actually interested in what you say. Stan does operate on his own time, which often slows things down, especially on the golf course. Sometimes it's necessary to say, "just hit the fucking ball". But otherwise, I'll always pick Stan on my team.

As you can see, I'm fond of my golfing buddies. We all couldn't be president of M&E, so I guess it's natural that many of us left. But the time we worked together, at the dawn of this country's environmental awareness, was special for its energy and hope. We were all lucky to have that positive experience, that helped form the rest of our lives, for the better.

CHAPTER 4

RIZZO ASSOCIATES, 1985-1999

When I left M&E I didn't have another job waiting. I was seriously thinking about somehow getting into the real estate development business. Ever since working for a mason, building houses after my freshman year in college, I was attracted to small scale development. The idea was to buy rundown buildings in Boston, renovate them and resell them, hopefully for a profit.

By way of background, I was lucky to have bought 69 Mt. Vernon Street, but after Erin was born, Cynthia wanted to move to the suburbs. To make money for the move, we sold the building. I should have made it into condos and only sold one unit. It would have been more than enough for a down payment. Big mistake, but the sale did generate enough to buy a house in Wellesley Hills, with $400,000 left over.

For tax purposes, I planned on an exchange. At the time, this allowed you to buy a "like kind" property(s) within 3 years and avoid the capital gains tax. A very good deal, but I needed to invest the $400K in one or more residential properties to make it work. This was part of the reason I was thinking about going into the real estate development business. I eventually bought a building in Jamaica Plain for cash, so I ended up with two houses and some money left over. It was a conservative approach, but stupid from a real estate perspective. I should not have sold the Beacon Hill house, but once I did, I should have leveraged the money and bought several buildings. This would have entailed down payments and large mortgages, but in hindsight, given the appreciation in Boston real estate, it would have been a wise approach.

The plain truth is I didn't have the risk tolerance to assume a large amount of debt, including mortgages and construction loans. There would have been some rent income, but if you are renovating a building, there is no rent. Not having a day job further elevated the risk. Cynthia was also not working at this time. The end of my real estate career, which never really started, was a call from Bill Rizzo.

Bill and I worked at M&E, but only had one interaction which didn't go well. I was the project manager on a large project in Boston that involved improving the water quality in the Back Bay Fens and the Charles River for storm water runoff and combined sewer overflows. As part of the improvements, we wanted to make significant modifications to a large stone gate house on the banks of the Fens, across from the Museum of Fine Arts (MFA).

Bill was in charge of preparing the environmental assessment (EA) for all the improvements we were proposing. Bill came to me and said we couldn't change the gate house because the architect was H. H. Richardson, and it was a protected historic building, built around 1900. H. H. designed Trinity Church, among other famous buildings, and had quite a reputation. I said that was silly. To improve the water quality in the Back Bay Fens and the Charles River we needed to rearrange some of his famous stones. I was a bit forceful in my comments. Bill was having none of my arguments and we had a standoff. He said his hands were tied, and I knew I was going to eventually lose the fight. After several days of me pouting, I agreed to move our equipment to a new building and leave H. H.'s alone. In the end the project was never built, but we always laughed about how we met.

Bill had left M&E to become the assistant secretary of transportation in the Executive Office of Transportation in the administration of Governor Ed King. Bill knew the secretary, Jim Carlin, because his sister was a college friend of Jim's wife. In any event, Bill loved his job and, as a result, got to know Jim better. Sadly for Bill, Mike Dukakis came back to beat Ed King after his first term. When Dukakis' people came back to power, it wasn't long before Jim, and then Bill, were shown the door.

Jim had a very successful insurance business, so he had a job to go back to, but Bill didn't. Jim was partly responsible for this, since he hired Bill, so he proposed that they start an engineering consulting business, each owning 50%. Jim would provide startup capital and Bill would provide the sweat equity. And so, Rizzo Associates was born in 1983. Jim's insurance company was in a building he owned in Natick, so he provided Bill with space in the building.

Bill had a friend at M&E, Liz Levin, who I also knew, but not well. They kept in touch and when Bill started Rizzo Associates, he told Liz he was hiring. When I left, Liz mentioned it to Bill, and I soon got a call. At this point I was still evaluating the real estate adventure and was not ready to take a full-time job. We agreed that I would help on a few projects as a part-time consultant. After a few months, I was working full time, and since the real estate stuff wasn't working, I became one of 7 employees: Bill, two admin folks, two engineers, one geologist, and me. I was the second employee with a professional engineering (PE) license. Bill wasn't a PE, but had a BS and MS in civil engineering and was a city planner in his early days.

For my entire career, up to this point, whenever I made a phone call, I would start by saying, "I'm Rick Moore with Metcalf &Eddy". Everyone in the business knew M&E, and you were proud to be part of a company with such a history and reputation. Now I had to say, "I'm Rick Moore with Rizzo Associates". Who the hell knew Rizzo Associates. There were times I had a hard time getting it out.

The good news was my salary was much improved over M&E, my title was VP, and I had a good boss. All indications pointed to Bill being a quality guy, but I had little actual firsthand proof. After all we didn't agree about the environmental assessment described above, and to make matters worse, he went to Lehigh and I went to Lafayette, arch rivals. The matter was settled for me a few years later when I had testicular cancer, with 4 months off for chemotherapy. Bill, without fuss or discussion, continued to pay my full salary through this difficult period, when the company had little money to spare. You never forget something like that, and it turned out to be only one of many instances that showed Bill's honorable character.

Rizzo Associates started on the strength of Bill and Jim's connections, even though their approach to business was very different. Bill was low key and, somewhat naively, thought you should be selected for a job based on qualifications. Of course, it helped if you had a close personal connection with the person making the selection. He bent over backwards to avoid even the hint of a conflict of interest. He would say we don't want to wake up some morning and see our name (his name) on the front page of the Globe, above the fold, involved in a scandal. We actually lost some jobs because Bill played it so straight.

Jim, on the other hand, was more than willing to push the line and make a phone call when necessary. His approach was, your friends do you favors, and you do favors for your friends. At no time was I aware of anything improper, but I wasn't a part of Jim's calls. Two other things were a bit frustrating. First, our competition just assumed Bill, and especially Jim, were calling in favors. After all, Jim was the previous secretary of transportation and he gave a lot of money to both Democrats and Republicans. So, even if we beat them fair and square, they felt the fix was in for Rizzo Associates. Second was Jim's friends. Fortunately, we didn't do much work for them, but when we did, it usually didn't go well.

One more thing about Bill's style of management. He was a great boss with high standards and a hands-off approach. He was more interested in doing the right thing than making money. I think this came from his time in the public sector. Early in his career he was town planner in Leominster and later assistant secretary of transportation. He loved these jobs, and I think his nature was more attuned to the public sector than the private sector. This was the complete opposite of Jim.

Along these lines, Bill was friendly with Gerry Kavanagh, the city planner in Salem, MA, who went on to be Ted Kennedy's chief of staff. A developer, David Ebershoff, from Cleveland, Ohio, had a project in Salem and asked Jerry who would be a good firm to help with permitting and engineering. He gave Bill's name. The timing was good, because that was one of the reasons Bill called me about work. He needed someone to manage the Ebershoff project, both the technical part and the political/permitting part. Dave was a character, 6 foot 8 inches tall and a retired NBA basketball player. At an early meeting with town officials, he suddenly did a rendition of the Music Man song, "Ya Got Trouble" in River

City. He was quite good and, strangely, it made a good impression with the town.

I got to know Jerry and, with his help, got Dave's project permitted and designed in short order. Jerry also went to Lafayette, which showed that Bill didn't hold that against you. Everyone was happy and I got off to a good start. Jerry said I had written the best permit application he had seen and was going to use it as a template for other developers. M&E taught me well!

One other Jerry Kavanagh story, that happened much later. Bill was involved in pursuing work in Poland and, as part of the effort, we supported a visit for a small contingent of Polish officials to Washington, DC. At the time Jerry was Ted Kennedy's chief of staff, so Bill arranged, through Jerry, for the contingent to meet the Senator. Given that Poland is 95% Catholic and that JFK was right up there in reverence with Pope John Paul II, the meeting was a big hit that very much impressed the Polish folks. Afterward, Bill asked Jerry if there was something we could do to thank Kennedy. Jerry laughed out loud and said of course, there were three options: money, money, or money, the mother's milk of political life. So, we held a fundraiser at the Engineering Center and did just that.

By the time Dave's project in Salem started construction in the Spring of 1984, I was a full-time employee. With the work in Salem done, I needed to find something else to do, since there weren't many other jobs in the office. The ones we had were in two general categories. First were private sector jobs for small time developers, like the one in Salem, and the second were hazardous waste investigations.

The good news for Bill was that the State had just passed a hazardous waste bill. In many ways, it followed the federal superfund law. Because of cancer concerns at Love Canal in up-state New York, WR Grace in Woburn, MA (A Civil Action), and other scary stories, the public was up in arms about hazardous waste contamination.

The Massachusetts Law (Chapter 21E) had essentially created a new industry by requiring that the buyer of a property, other than a single family home, accept responsibility for any hazardous wastes/materials on the site. Even if a developer wasn't worried about this responsibility, the banks were, and wouldn't lend money unless the developer did a hazardous waste site assessment as part of normal due diligence. Who better to do a site assessment, than Rizzo Associates. Hence, a brand-new market was born.

Many companies, including M&E, didn't want to get into the hazardous waste business because of the risk. The hazardous waste due diligence business basically had three steps. First was a desktop assessment looking at available information (e.g., location of underground gas tanks) and making an educated guess about the likelihood of pollution. If the answer was yes, wells were driven and samples taken to see if the soil and/or groundwater were contaminated. If that was another yes, a clean-up followed, usually at great expense.

The risk was saying no to pollution, after the first step. If the developer bought the property, in part based on this determination, and found pollution while digging a foundation, the lawsuits started to fly. This happened even if your contract had extensive language about "limits of liability". When the industry was getting started in the 1980s, the uncertain legal landscape scared most of the big firms away. One lawsuit could put you out of business. Bill took advantage of this opening and went after "21E" work as a major opportunity, as long as he was careful. More on this later.

Back to the private sector development work, which was more to my liking. Keep in mind that I had never done private sector work. M&E had all public clients, as they believed you couldn't do both public and private work since they were so different in their demands. They also were concerned about getting paid. If a developer approached M&E with private work, they required a 100% retainer, which no developer was willing to accept. Hence, no private work. After I did the Ebershoff project in Salem, I thought M&E's approach was faulty. You could do both, although they were right about getting paid. A lesson I learned quickly.

The advantages of private work, although it may not be as interesting as public work, are the money and the lack of bureaucratic bullshit. But mostly the money! If you have relationships with quality developers, you get repeat work without competing and you get paid well. Of course, your relationship and getting paid are based on delivering the required outcome on time. It's high pressure, but I thought a good bargain. This was especially true for me since, although I had good contacts in the public sector (City of Boston, MWRA), it takes 9 to 12 months to win a job, negotiate a contract, and start work. I didn't have that time, and with a staff of two people, there was no way a public client was going to hire us. With a private client, you could get a call one day and start work the next.

Again, I was lucky in that Jim and Bill had some good private sector connections. The one that helped the most was with Finard & Company, developers of Natick Mall. They wanted to re-develop and vastly expand the Mall, into one of the biggest in New England. The fact that our office was in Natick, 15 minutes from the Mall, helped us get the job to do preliminary engineering and permitting. Of course, Bill's connection with George Khouri of Finard and Jim's connections in Natick sealed the deal. It was a huge job for a small company and, as project manager, it kept me busy for months. This also helped while I developed my own private sector clients. I liked this work, and my M&E experience taught me to do things right.

Over the years we did many of these types of projects, all of which generally followed the same pattern. You needed to get state and local permits, and to get them you needed to do some preliminary engineering. Typically, first you passed the state environmental review process, which was subject to the Massachusetts Environmental Policy Act (MEPA), and required the preparation of environmental impact reports (EIRs). For big projects, these could be several volumes, describing the impact and mitigation for traffic, utilities, wetlands, and any other issue judged by the State to be significant (e.g. air, noise, wind, shadow, historic preservation).

Once you passed the state review, it was on to the local city councils, selectmen, planning boards, conservation commissions, and others. It was a long and expensive process, with lots of writing and meetings. Fortunately, my time at M&E prepared me for the writing. The meetings required formal presentations and I was usually well-prepared. I liked the give and take with the local boards and commissions, despite the many, many nights out.

A word about presentations. I was fortunate that the Army taught me to teach, and then assigned me to the Army Engineering School at Fort Belvoir, VA, for one year. This was valuable training for talking in front of clients, Boards of Selectmen, or presenting a professional paper in front of 50 other engineers. I learned three important rules while teaching. First, practice, practice and practice some more. This makes the next two rules easier. Second, know well how to transition from one topic to another. This makes for a talk that flows, without awkward dead spots. Third, don't read your talk. Talk to the audience and make eye contact. You can't do this if you forget the first rule.

My favorite story about talks/speeches, comes from Douglas MacArthur. When President Truman relieved General MacArthur from command in Korea, he came back to the US a hero, after being overseas for over 10 years. Truman was right to relieve MacArthur, but many didn't think so. MacArthur was thinking of running for president and went to West Point, where he graduated first in his class, for a visit. He was at lunch with the cadets, and many newspaper folks, when the Superintendent asked if he would say a few words. MacArthur appeared as if this was a big surprise, but he had prepared a speech and memorized it. It had the famous line 'the long gray line has never failed us', among others. The Press went crazy, thinking that he had come up with this amazing speech, off the cuff. He had prepared, practiced and spoke to his audience.

Some of our biggest projects, requiring endless meetings and presentations, were the Natick Mall, South Shore Plaza, Patriots Stadium, Mohegan Sun Casino, Westwood Station, several transit oriented (MBTA) developments, and many, many plain vanilla residential, commercial, and office developments. Over the years, we also had several clients with multiple projects going on at the same time (e.g., Marriott Hotels, Lowes, Postal Service). These were great clients, with lots of work that kept many people busy at good billing rates.

In order to do these projects successfully, Bill had a plan, which I agreed with. We would be the lead consultant, performing work in several key disciplines, and subcontract the more exotic issues, like air and noise. The key disciplines were site/civil engineering (utilities, grading, and drainage), traffic, wetlands, and hazardous waste assessments. When I started, we only had site/civil engineering (me), so we had a lot of subcontractors. We needed to build a much stronger in-house team, and we needed to start with the lead person in each discipline.

4A. GOOD PEOPLE.

Bill was very good at attracting quality people. After all, he convinced me to work for him in what was essentially a start-up business. He soon attracted two more good people, Andy Magee and Rick Hughto. Andy was a geologist, so knew his way around all things environmental, and was an excellent writer. He worked for me, primarily putting together environmental impact reports (EIRs) and was a pleasure to work with. Rick Hughto was a PhD from Cornell, who was a hazardous waste expert. He came from CDM and was the best of the best. Like me, Rick came from a large well-known firm to a very small one. He checked out both Bill and me with several mutual friends and found us competent and of good character, so signed up. Because of his exceptional credentials and the fact that the hazardous waste business was taking off, he headed his own group, reporting directly to Bill.

Around the same time, I needed a civil engineer to help me, so I reached out to Brian Sullivan, who I knew from M&E, and he came on board. Brian was a good Irishman, who could sell you the Brooklyn Bridge, but was at heart, and underneath the big smile, a conservative engineer. We did many projects together, often working long hours to get things right. He grew in all respects related to technical and management issues and eventually took over my job(s) in the company. My main complaint, was, and remains, his inability to be on time. He missed the start of more than a few meetings and once we missed a plane to NY City for a client meeting. I would go bullshit, but it didn't do any good and, because he was so good at everything else, I let it pass.

It took some time to get a lead traffic engineer, but after several mistakes, Bill hired Barry Pell. Barry danced to his own drummer but was a gifted traffic guy. He was not interested in administration, nor did he have a great bedside manner, but clients loved him for his exceptional understanding of the technical traffic stuff. Barry, unlike many traffic engineers, could also write well and speak in plain English. I found him quite engaging, since he was a renaissance man interested in many things besides traffic.

To illustrate Barry's independent streak, he took on a job representing a neighborhood association fighting a retail development in their midst. It was a favor for a lawyer he knew, who happened to live in the neighborhood. He didn't mention it to us, because he knew we wouldn't like him fighting a developer, even though he didn't know who the developer was. Barry wrote a blistering indictment of the traffic impacts that was submitted to the town in opposition to the development. Since we, including Barry, always represented the developer, it wasn't good form to take the other side.

This lapse of judgement would have ended there, but a month later Bill and I took a trip to Bentonville, Arkansas to market Walmart for work in New England. As you can imagine, this was a big deal, with a huge potential client. Before we could even make our pitch, their chief development guy tossed a copy of Barry's report on the table and asked us to explain. It had cost them a lot of heartburn rebutting Barry's arguments against their project. Now we knew who the developer was, and even though they acknowledged Barry's brilliance, we never got any work from them.

All four of these guys, Andy, Rick, Brian, and Barry were professionals at the top of their game. They could talk confidently to any client, convincing them they knew what they were talking about and also go head-to-head with any regulator or local official to make their point and well represent our clients. They had good business sense, especially Rick and Brian, and this kept us out of most lawsuits for the life of Rizzo Associates. It's hard to put a value on something that didn't happen, but we all knew that a serious lawsuit could cripple the company. In short, these were good people to associate with.

The company was divided into two technical groups. One was hazardous waste, led by Rick Hughto, and the second was everything else, headed by me. But there was one more important addition, again from

M&E. You think there was a pattern here?! Bill hired Liz Levin to be our marketing director. Liz was what I called a "senior woman" in the civil/environmental engineering consulting world. It had nothing to do with chronological age. When we entered the work force in the 1970s, there weren't many women in the mix, and most were in the public sector. Liz was one of the few senior women, a planner, not an engineer, who was in the private sector. She had cut her teeth over about 10 years in a very competitive environment at M&E. She was clearly an accomplished professional, Bill liked her, and we needed a marketing director. I also knew Liz, although not as well as Bill, and thought his decision to hire her was a good one. Bill was also big on supporting women, and this was clearly a step in that direction.

After a few years, Bill diluted the ownership stakes by gifting Liz, Rick, and me a minority ownership stake. This was clearly not Jim's idea, but Bill argued that this needed to be done, or he would lose us, especially Rick. Jim finally agreed. So essentially, the four of us formed the executive team, with Jim as chairman of the board. Bill and Liz were clearly attuned to collaboration, group decision-making, and happy endings. We can call them the Kumbaya group. Jim, Rick, and I (often called the two Ricks) were the opposite, "make a fucking decision and move on." I was more in the middle of these two management approaches, but Jim and Rick were clearly top down decision makers.

At about this time, Bill and Jim sold shares to form a 40% ESOP (Employee Stock Option Plan), enabling all employees to indirectly become owners. This was consistent with Bill's philosophy, but Jim came along kicking and screaming.

It was good to have different opinions. Jim and Rick were all about the money and Bill and Liz put value in other things, like social responsibility and maintaining good ties to the outside world, which often included making financial contributions to good causes. They also wanted to treat employees and subcontractors well, in Liz's case often too well. As marketing director, Liz provided an endless supply of opportunities to spend money on good causes and marketing materials (e.g., company brochures and sponsorships).

Liz's efforts clearly improved our 'brand' and took the long view. However, with no immediate results (signed contracts), this took away from the bottom line and drove Rick crazy. Rick finally hired his own marketing person and would have nothing to do with Liz. So not only did we have two technical groups, but we essentially had two companies reporting to Bill, and Liz had no role in Rick's side of the business.

It was actually better to do this, rather than having constant conflict between the two. Rick was excellent at managing his side of the business, and the money was so good, Bill wasn't about to shove Liz down his throat. I could see both sides and had good relations with both of them. My side of the business, which was in part tied to local developers, was much more cyclical and not as profitable as Rick's. Given this, Rick had a higher standing with Bill than I did, for good reason.

An example will illustrate this point. During my time at Rizzo Associates and Tetra Tech, there were three serious economic downturns. One in 1988-89, when the "Massachusetts Miracle" collapsed, due mostly to bank failures; one in 2000 when the dot-com bubble exploded; and the last in 2008, when real estate mortgages went bad. All of these dramatically impacted our private sector market in a negative way. So, let's focus on the 1988-89 downturn, which caused the development market to shut down. Our best client at the time was Marriott Hotels. We were working on 8 to 10 projects throughout New England in various stages of development, from initial due diligence through permitting, design, and construction services. One day in 1988, I got a call to "put down your pencils". All projects were being put on hold, except one or two under construction.

The Marriott call was a bad day, as we had to lay off a few people, since all our private development work was slowing down or stopping. This slowdown also hammered the hazardous waste business, but for one significant positive change. Because so many developers were defaulting on their loans, the banks were taking back hundreds of properties. By doing this, the banks became responsible for any hazardous waste clean-up that was necessary on the site. To limit their liability, the banks wanted new site assessments done on all the properties they were considering taking back. This was a bonanza for Rick Hughto and his merry band.

It's not an understatement to say that Rick's work for the banks, and the lawyers that represented the banks, saved Rizzo Associates from a severe financial crisis in 1990. Jim also had a real estate business that almost went under at this time and shook his whole business empire to its foundation. It may sound funny, but the one thing that most illustrated the dire straits was when Jim let his driver go (a retired cop from the Natick police force), and had to get back behind the wheel. Fortunately, the black clouds passed by the end of 1991 and the rest of the decade was a glorious run for the roses.

This illustrates, and I'll say it again, how important Rick Hughto and the hazardous waste business were to the financial stability of the company. Bill understood this and valued Rick accordingly, as was only fair.

Now let's talk about the side of the business I ran, which covered so many bases it was hard to keep track. This was both an advantage and a disadvantage from a business point of view. The advantage, at least in theory, was that, if one segment went soft, another could take its place. If private sector work slowed, think 1988, public sector work could expand. A disadvantage was getting enough strength (e.g. staff) and reputation, in any one area, to be a force in the market.

After some time, we had several disciplines in-house, including transportation (traffic and highway), environmental (wetlands, water, and wastewater) and site-civil (drainage and grading). We sold these services to both public and private clients, so there were always a lot of balls in the air. Engineering firms with several disciplines always have the dilemma about how to organize and sell their services. Should it be by discipline or client. Regardless of how you craft an organization chart, it's really a bit of both approaches, and depends largely on collaboration among the senior people.

Most firms favor organizing around disciplines, such as a highway department with lots of highway engineers selling highway design services to a transportation agency. That's simple but doesn't leverage "cross selling" other services to the same client. The key is if the highway person is willing to look for opportunities in other disciplines. For example, we did transportation, environmental, and hazardous waste work for the Massachusetts Highway Department and Massport, because Bill, Rick and I saw the benefit of this cross selling approach. Problems occur if the senior person managing a client is concerned that bringing in another discipline will, in some way, damage his/her relationship with the client. It's called protecting your turf, and is not an enlightened path to business success.

Since I had a long history in the public sector at M&E, I tended to take the lead in these types of projects. We did a lot of work for the Boston Water & Sewer Commission, MWRA, Mass Highway, Massport, and the Postal Service. Brian focused almost exclusively on private sector work for Marriott, Lowes, Cabot, Cabot & Forbes (CC&F), New England Development, The Chiofaro Companies, and many more. The goal was to sell all our services to all our clients. For example, Bill brought in the first Postal Service job, partly because his father owned several buildings where they were tenants. It exploded into so much work for Rick and I, that they were our biggest client for several years. In another case, Rick had a relationship with Jay Doherty, President of CC&F, and he introduced Jay to me. We then did half a dozen large transit-oriented jobs near MBTA/Amtrak train stations for CC&F. I had a long-term relationship with Len Liguori, who became the lead person for Lowes in New England, much to our advantage.

Len was a new Lowes employee, the only one in New England. Lowe's started in the south, with headquarters in North Carolina. They couldn't believe the complicated process of getting a project permitted and built in New England. In many places in the south there wasn't zoning, and you could build without permits. They wanted to compete with Home Depot in New England, but thought the process too long (1-2 years) and too expensive, north of the Mason-Dixon line. I agreed with Len that we would do some free work up-front to help identify good sites, get corporate interested, and also make Len look good with the home office. Eventually, Brian and Rick did a huge amount of work for Len on dozens of sites, as Lowes penetrated the New England market. It didn't hurt that after the first Lowes was opened in Massachusetts, the sales were 4 times higher than in similar stores in southern states. It was worth the cost and waiting for the permits, so we were off to the races.

Over the period from the early 1980s to mid 1990, the company steadily grew to about 200 people, with annual revenue of about $20 million. We had become one of the top consultants for private developers in Massachusetts and a mid-level player in the public sector, especially for state agencies. Our reputation was excellent, and we started to feel like an important player, if not at the same level as an M&E. There were, however, two internal issues that weren't apparent to the outside world, but that wouldn't go away. One had to do with Jim Carlin and the other with Liz Levin.

Jim was a successful businessman with a very strong, street-smart personality. He was also well connected politically, which was true because of his good-old-boy personality and the fact that he gave generously to both political parties. Aside from the mini recession in the late 1980s Jim had made a lot of money in his insurance and real-estate businesses. In addition to starting/funding Rizzo Associates, he started several other businesses related to us. One, a chemistry laboratory, and the other a drilling company, both supported by our hazardous waste business. He also helped us buy a small survey company, a small transportation company and a small engineering company, with yet another former M&E legend, Dave Duncan. The lab was the only one that you could call a success, but Jim was a real entrepreneur and risk taker.

Jim also liked awards and had Bill submit our company statistics to BusinessWeek magazine to qualify as one of the 500 fastest growing businesses in the U.S. We made the list several years in a row. It was fun, but not too hard to have high statistical growth when you are a small company. If your annual revenue grows from $1 million to $2 million, that's a 100% growth rate, but you are still a very small company. Jim had a plaque he could hang on the wall and show all his friends. I did too. We also, after some time, made the Engineering News Record's (ENR) list of 500 largest US architectural/engineering (AE) firms. This was a big deal. ENR was the Bible of the industry, and, to me, tangible recognition that we had arrived, no longer in the shadow of firms like M&E.

One of my favorite stories, which best illustrates Jim's connections, occurred at the end of Mike Dukakis's second term as governor, when he was running for president. Even though Jim had worked for Governor King, a Republican, he was close to Dukakis and wrote him big checks. In Massachusetts, conservative Democrats and liberal Republican were close to the same, and, in any event, in business you needed to butter both sides of the bread. To illustrate the connection, Dukakis named Jim the receiver of the city of Chelsea when it went bankrupt. A big deal, but I digress.

There was a time around the Democratic convention when Dukakis was 20 points ahead of Bush in the polls and he seemed a sure thing. Having Jim, with the ability to call the president on the phone, and get a return call, was a clear business advantage. Beyond that, Dukakis was considering Bob Gramm, senator from Florida, as his VP. It just so happened that Jim and Bob were college roommates at the University of Florida, and still close friends. How good would that be!! We all know that Bob Gramm wasn't picked, and The Duke lost, but it was fun to think of the possibilities, even for a short time.

Jim started an investment group for the senior people in each of his companies. The plan was for each company to match each dollar you invested, so you immediately had a 100% return. Very nice benefit. The money was to be invested by Jim in high risk/high reward ventures, so, as the saying goes, don't invest the money you need to send your kids to college. Over several years, investments were made in radio stations, real estate, and banks. It was a great idea until it wasn't, when the 1989 recession hit. The fund lost about 50% of its value, but at least we got our original investment back. The cruel part was that right after the fund was dissolved, Jim and his close friend Jim Kerasotes helped start a company with Jeff Taylor, called Monster.com. We all had a chance to get in at the beginning, and Jeff seemed like a capable guy, but based on recent experience, passed. It was sold about 5 years later for over 100 million dollars and is still in business. So goes life in the fast lane.

This was all fine and having Jim as a financial safety net was wonderful, but his view of life was very different from Bill's. It eventually caused the sale of the company, but during the early years there were signs of trouble ahead. These often came up at board meetings, which were attended by Jim, Bill, Rick, Liz, and me. We had an outside board member from time to time, but they were not consequential. Jim was far more financially driven than Bill and was always complaining about our financials, and he was right. We were doing everything right otherwise: growing, creating equity, developing an excellent reputation, doing good in professional organizations, and taking care of our employees. All these things were good business practices, and pay off handsomely in the short and long term. What we never did to Jim's satisfaction was make enough money.

To summarize, the enduring conflict was that Jim wanted to pull money out of the company and Bill wanted to reinvest it in growing the business and rewarding the staff. Establishing the ESOP was Bills major success on his side of the divide.

We had a lot of debt and never enough cash to give serious bonuses, especially to Jim. At one board meeting during the Massachusetts Miracle in the late 1980s, when every company was making money, Bill was so proud to present our latest financial results. After he finished, Jim blurted out that any idiot could make money in the current environment, or words to that effect. This was at a board meeting. God knows what he said to Bill privately, and he said a lot.

Jim said what he thought, and the tension was always present. Bill mostly protected us from Jim but couldn't protect himself. In short, Bill was a professional who wanted to do good and made decisions consistent with his philosophy. Jim was a businessman who wanted to make money, and made decisions consistent with his philosophy. Rick was clearly in Jim's camp, Liz was clearly in Bill's camp, and I was in between. Bill and Jim had to agree on any important issue, so you can see that this match was not made in heaven.

We never had a large bonus pool at the end of the year, and some years we had to borrow money to pay even modest ones. One year, instead of giving many small bonuses, we decided to give one significant one, $10,000, which was still small by most standards. The lucky person was Paul Cincotta. Paul was and is a terrific guy, who I still go fishing with. At the time, mid 1990s, he was a project manager and we had won a large project for the Town of Wayland to redesign many miles of rural two lane roads. I don't remember who was the first project manager, but that's just as well, because the job was way behind schedule and over budget. We asked Paul to step in and salvage the project and our reputation. There was a committee set up by the town to oversee the project and they were chewing us a new ass hole.

Paul took over and got the job done, minimizing our losses and repairing our reputation. It was a thankless job and Paul did it with good humor. His bonus was well deserved. This was the kind of thing Bill was good at, looking behind the numbers and identifying value. We agreed it was the right bonus decision. It wasn't much money, but it sent the right message, even though there were some project managers who were successful financially, and thought otherwise.

Liz was a different kettle of fish. As mentioned, Bill and I knew Liz from M&E days and were very fond of her. She was very smart and good at analyzing problems. She went to Wellesley, had an MBA from Simmons, and a Masters in Urban Planning from BU. Politically, she was a very liberal Democrat. She definitely didn't travel in the same lane as Jim, and I suspect he thought she was a bit of a space shot. She was our marketing director, but only worked with me. Rick Hughto would have nothing to do with her. Enough to say that they were very different people: Liz a "pointed headed" liberal and Rick a "greed is good" conservative. Fortunately, Bill played referee very successfully for many years.

As marketing director, Liz had three main responsibilities. First was the preparation of company information (e.g., brochures, project write-ups, resumes) and the second was using these in the preparation of proposals. These were internal processes done in collaboration with senior staff and project managers. The third area had an outside focus and consisted of gathering intelligence about potential future projects and our interaction with professional organizations. She had a staff of two people who helped with company information and proposal preparation. What she most enjoyed was interacting with the outside world by promoting Rizzo Associates.

Keep in mind that she had no involvement with Rick Hughto in these areas. He hired a separate marketing person for his business, who didn't see eye to eye with Liz. She also had no significant contacts with the private sector, think developers. Liz was really working for my side of the business in the public realm. Given her salary, the fact that she was 100% non-billable and working for only about 25% of the company, there were some who questioned her usefulness. Nevertheless, she was effective in promoting the company to the outside world. She was a senior woman in our business universe, and a highly respected member, often president, of a number of professional organizations. She knew many senior women, and many were high up in public agencies, like Massport.

Two examples will illustrate my point. We were going after a large contract with Massport, to prepare and negotiate their permit allowing discharges of storm water and deicing fluids from Logan Airport to Boston Harbor. It was a big deal, and all the major firms were competing with great vigor. We put a good proposal together and had an excellent interview. The key decisionmaker was the head of environmental affairs for Massport, Robin Ellis. Liz knew Robin and told her she would like working with me. This was key in making Robin comfortable with me, and we got the job. Clearly, Liz spread much good cheer among the women in our profession, but there weren't many examples, like Massport, where the goodwill converted to a signed contract.

On the other side, we had a significant opportunity to get work with a major NY shopping center developer, in this case the redevelopment of South Shore Plaza in Braintree, MA. We had an interview where Bill, Liz, and I made our pitch, with the key that Liz would be the day-to-day project manager. We thought they would like a woman as their main advocate before the town of Braintree, and it would also get Liz some billable work. Unfortunately, during the interview Liz overdid it and sounded more like she was working for the town. After the interview, their key guy, Jamie Selonic, took Bill aside and said we could have the job, but he wanted Liz out and me as PM. It was an easy decision. We got the job, and I became good friends with Jamie for the next two years. He was a quality guy and one of the best, if not the best, client we ever had.

As you can see, Liz could be a two-edged sword. I was friends with Liz and supported her until the end. Bill also thought of Liz as a major advantage as the face of Rizzo Associates to the outside world. So few consulting firms had senior women that this was a competitive advantage for us, especially since most senior women were in the public sector and were potential clients. This all worked, sort of, until Bill created a significant change in our organization in the late 1990s. He proposed that Liz become head of our transportation group. This made some sense, since Liz had experience with the transportation agencies, think Robin Ellis at Massport, and it was an opportunity to get her billable. Her salary was all going to overhead, a big drag on our profitability, which made Rick Hughto go crazy. Me too, sometimes.

This reorganization was taking away some of my responsibilities but I wasn't a transportation guy, and it made sense. I was fine, actually in some ways happy, with the change. I'm not so sure about the rank-and-file transportation guys, and they were mostly guys. That was a manageable issue, but a bigger problem developed because Liz was to give up her role as marketing director. I assumed, as I think most people did, that she would recommend her loyal assistant of many years, Mary Barry, who in my opinion had earned the promotion.

For reasons that I still don't understand, Liz instead picked a mid-level geologist in the hazardous waste group without any marketing experience, to be her replacement, with Mary reporting to him. This seemed crazy to me, and I said as much to Bill, but he wouldn't overturn Liz's decision. Here was a so-called champion of women, who proceeded to humiliate one in front of the whole company. Perhaps Mary wasn't the perfect lead person in Liz's eyes, but who was. She went on to run CDM's marketing department for many years, a firm 10 times the size of Rizzo. Liz's choice, Matt Marshall, was a terrific, talented guy, who could have been Mary's assistant (maybe boss someday), but wasn't ready for the top job. I still get upset thinking about how wrong that was, and my relationship with Liz was never the same.

When speaking of transportation, there is no way to avoid mentioning the Central Artery/Third Harbor Tunnel project, affectionally called the Big Dig. Although it often gets a bad rap because of its cost, $15 billion, it was a classic example of a public investment that, together with cleaning up Boston Harbor, transformed the city. These two public projects turned out to be worth every penny that was spent, and more. Without them Boston would not be the world class city it is today. Of course, these projects created a feeding frenzy for consultants, and we wanted in on the fun.

This was a natural for Bill, given his experience in the Department of Transportation, but first we needed a credible project manager, who was respected by the Highway Department. Bill hired Harold Fitzgerald, who had been around the block a few times and was not being treated well in his current job. Fitzy was well-known at the Highway Department, so we could build a team around him. He was also a wonderful, decent guy.

The Big Dig had a management consultant, Bechtel-Parsons, who did the permitting and preliminary design for the whole project. It was then divided into sections, and consultant teams bid on each section to complete the design. Once completed, contractors could bid for the work and if successful, build it. We were successful on two sections, the southern most portion of the Artery, and the northern most section, mostly on Logan Airport. The southern section was a highway viaduct between the South End and South Boston. We were the prime consultant, with two major subcontractors, and Fitzy was the project manager. The northern section also consisted of viaducts, that connected the Ted Williams tunnel with Route 1A North (East Boston Expressway). We were a subcontractor to the prime on that project.

These were good projects, and we made money despite the restrictions the Highway Department had on overhead rates and salaries. The way this worked was by setting up a project office with about 30-40 people working only on the project, with 40 hours or more of billable time each week. For the southern project we had an office near South Station, and we accounted for about half of the total staff. Besides salaries, you charged everything else to the Highway Department, rent, utilities, supplies, everything! This kept the overhead for the project very low, and as a result you could make some money.

We did have one close call on the southern section. The roadways were supported by large steel box girders that were welded together in the field. As they were being completed, we discovered that some of the welds were smaller than called for in the drawings. The good news is that we discovered the problem, and after alerting the Highway Department, did some structural analysis that concluded the welds were OK despite their smaller size. I lost some sleep over that, and so did Fitzy, but it ended well, and the Highway Department thanked Fitzy for dealing with the problem as he did.

The northern project at Logan Airport progressed without major issues. There was a major non-highway component to each of the contracts. Each included a large storm water pumping station. For each project, all the runoff was collected by gravity at a low point and then need to be pumped to Boston Harbor. The stations were designed for the 100-year storm, so they were huge structures with giant pumps, which sat silent most of the time, until it rained. When it did rain, the pumps were needed, or the area would flood. This was not a simple design.

The person who stepped up and managed the designs for both pumping stations was Dave Duncan. The designs were very complicated, and Dave guided them to success, while teaching the younger engineers how to do things the right way. Dave was a senior VP at M&E when I was there, left to start his own company. We bought his company and made him our chief engineer. He was an engineer's-engineer and although past retirement age, loved his work.

When you first met Dave, you were totally intimidated, largely because he had a poker face that was impossible to read. You weren't sure if he was thinking you were an idiot or a genius. But once you scraped the surface, you found he was a wonderful man. He loved to teach the younger folks, and we had lots of them. His initials were DAD, and the folks called him Dad. It was perfect and they loved him. He died while working in his mid 70s. We all went to his funeral and when the minister asked if anyone would like to say something, spontaneously, several young engineers, men and women, stood up and said, in different ways, what a great mentor he was. A fitting tribute. I had tears in my eyes.

I was really blessed with the good caliber of people I worked with over the years, and many had their roots at M&E. One of those persons, who was outstanding, was Lisa Eggleston. M&E gave her an excellent start, she was smart, could write, and was good with clients. The only problem was she lived in Essex, on the North Shore, and had a miserable commute to Natick. This, combined with family issues (young kids), caused her to want to work only one day a week for a period of time. I was fine with this because I considered Lisa an exceptional person who I wanted to keep under any circumstance, even though it's hard to be productive one day a week. Two problems soon presented themselves.

The first had to do with raises. Lisa's position was that her raise should be based on the quality of her work and not discounted by the fact that she worked only one day a week. I disagreed for two reasons. First, Lisa was a project manager and really couldn't act in that capacity at one day a week. She was more of a technical specialist with much reduced billable time. I also had to explain, to the rest of the staff, why a person working one day was as important as someone working 5 days. Was that fair, and should there be some discount for getting such a flexible arrangement? The fact that she was a woman certainly made things more complicated, since women had long been given the short end of the stick in so many ways. I thought I was being fair to everyone and did discount Lisa's raise, which she didn't like. To this day, I'm not sure who was right.

The second case had to do with a technical paper. Lisa wrote the first draft on one about a project we had done for the Boston Water and Sewer Commission. Brian Sullivan also worked on the project, so Lisa sent the draft to Brian to edit, add the tables and figures, get the Commission's approval and get it ready for submission, things Lisa couldn't do. He did that and submitted the paper, with his name first, followed by Lisa's and the Commission's project manager.

Lisa was bullshit that her name wasn't first and I could understand why at some level. Lisa admitted that the paper would never have been submitted without Brian bringing it to the finish line, but she had done the first draft and felt she earned first billing. Since the paper was already submitted and I really didn't want to pick a fight with Brian, who wasn't willing to change, I told Lisa I wasn't going to force Brian to change. Lisa brought her argument to Bill and he said, in no uncertain terms, that a change needed to be made. It was, Lisa got first billing. This unfortunately had become a very contentious issue for all of us and it took some time for the feelings to cool. In hindsight, I believe Lisa was right and it was good that she pressed the issue. In this case, breaking the chain of command got the right outcome, but that was the part that bothered me the most.

4B. THE PROFESSION.

The idea of giving back to the community and profession was high on Bill's radar, and as illustrated above, also on Liz's. Liz was president of a major women's organization, the Women's Transportation Seminar (WTS), and Bill was president of the Engineering Center, an umbrella organization for the American Society of Civil Engineers (ASCE), American Council of Engineering Companies (ACEC), and MALSCE (Land Surveyors). I was active in the Water Environment Federation (WEF) on the national level, as chair of several committees over the years. I was also on the Board of ACEC Massachusetts and a member of the Boston Chapter of ASCE. Most of what we did was for the local chapters of these organizations, except my work for WEF. Bill and I served as chairs of our local conservation commissions for many years, Bill in Lincoln and I in Wellesley. Rick was not a joiner and stuck to business.

I enjoyed my work for WEF and made friends nationwide, meeting them at WEF headquarters in DC for meetings, or at their annual conference, mostly in New Orleans or Chicago. We collaborated on technical papers for their monthly magazine and had some fun. On a local level, our involvement kept us friendly with our competition, and we traded a lot of information both technically (ASCE) and business wise (ACEC). ACEC represented private consultants and was the lobbying group for our profession with the state house and public agencies. We wanted fair business practices and fought for them, mostly with the public sector.

In terms of business practices, there were two main areas that the consulting industry was constantly lobbying to change: procurement and billing rates. Of course, the public wants things done for the lowest cost, so competing for a job often boiled down to the lowest cost. Our argument was that picking the lowest cost engineer could lead to poor quality and much higher construction costs. Since the engineering costs are only 10-15% for a project, it's wise to get the best engineer to design the most cost-effective project. This was called Quality Based Selection (QBS), which promoted selecting an engineer based on their professional qualifications and approach to the project, then negotiating a fee for their services. Unfortunately, many public agencies had trouble explaining and selling this concept to the public or didn't try. The Boston Water and Sewer Commission had a chief engineer, John Sullivan, who understood QBS, and as a result they were wonderful to work with.

The second problem was with the limitations some agencies placed on billing rates. Engineers, like architects and lawyers, typically make money by billing hours. An hourly rate is made up of an employee's direct salary, the company's overhead rate (cost to run a business, like rent), and a profit. Towns, state agencies, and the federal government, all to a certain degree, audit and put limits on each of these components. The industry is always lobbying to eliminate or increase the limits. Sometimes you can't make money working for a particular client with very low rates, so you don't work for them. This fits the old saying of being "penny wise, but pound foolish."

The competition for work was intense, because like the hunter of old, if you didn't kill the deer (win the job), you didn't eat. This is where there was a major difference between the private and public sector. For the private sector it was much more "inside baseball", relying on relationships and handshakes. Once they were made you were in good shape and, if you worked hard, the relationship prospered with little traditional marketing. This was the case for many a good client, like CC&F, Lowes, Marriott, and Corporate Investors.

For the public sector, most job's were won only after a full-blown competition with sophisticated proposals, elaborate presentations, and detailed cost estimates. State-of-the-art marketing, graphic art, and IT departments were necessary supports. I was pretty good at putting proposals together and making presentations, so we got our share of wins, but the competition never let up. One time we lost a job in Rhode Island, because the fix was in for a small, local, well-connected firm. When I went for a debrief about why we lost, the executive director very proudly said that we had the best technical proposal, so he gave it to the firm that was selected and told them to do the work as we proposed. I should have called the FBI but decided against it and never did work in RI after that painful experience. That was an exception, as I never was approached about a bribe, or thought the fix was in with any other competition. Sure, large public agencies (Mass Highway, MWRA) wanted to spread the work around to qualified consultants, but that was just good business practice.

You might think that because of the competitive environment, the engineers involved were cutthroat, ruthless, and Type A personalities. You would be wrong. I'm not sure about other engineering professions (chemical, electrical, mechanical etc.), but civil engineers, in my experience, tend to be rather normal, quiet folks, often a bit boring. Good people. I think this is true across the breadth of civil engineering (environmental, transportation, structural, geo-technical etc.), but I'm mostly familiar with environmental water/wastewater engineers. That doesn't mean we don't work hard and play to win, but after the dust settles, we are ready for a cold beer and cordial conversation, knowing that the game was played fairly and will continue that way.

In some respects, I fit this description and in some I don't. I like a good fair competition and the cold beer after, but, like many engineers, I'm not detail-oriented and methodical. I prefer the big picture and broad strategies, leaving the details to others. I also have a bit of a temper, and can be head strong, which has gotten me in trouble, but when used sparingly can be an asset.

There were two large projects that we won which were especially gratifying to me. The first was an environmental project for the Mass Highway Department. Everyone is familiar with winter de-icing, where massive amounts of salt (sodium chloride) is applied to the roadways to keep them safe. Salt is a very effective de-icer and is cheap, but, as noted, it contains sodium and much of the sodium ends up being washed off the roadways into nearby surface and ground waters. In some cases, these waters serve as drinking water supplies, and high sodium in drinking water is a health hazard, linked to high blood pressure and heart disease. The national health standard for sodium in drinking water is 10 mg/l. If concentrations are higher, water utilities need to notify their customers.

One of the most heavily traveled roads in the Boston area is Route 128 (I-95), in Waltham. This section of a 6 lane Interstate highway, maintained by Mass Highway, is next to the Hobbs Brook reservoir, the drinking water supply for the City of Cambridge. The de-icing of 128, and other storm water runoff pollutants, were contaminating Cambridge's drinking water, so Mass Highway put out a request for proposal (RFP) to define the problem and come up with solutions. This was considered by the environmental consulting community as a signature project, because it was large (several million dollars) and covered issues that were common in many northern cities. The thinking was, if you won this project and did a good job, you could sell similar services to a potentially large market. Needless to say, the competition was intense.

We put together a good team, with Rizzo Associates in the lead and me as project manager. I got Mike Frimpter, who had just retired from the USGS, to be on our team. Mike had developed the "Frimpter method" to estimate high groundwater in Massachusetts and was considered something of a guru. He was a wonderful guy, like your friendly grandfather, and we took him to the interview for appearances. We won the job! The best part was the next day Bill and I got calls from M&E, who wanted the job badly, asking how we had pulled it off. It was so sweet, as we were now beating M&E, the best of the best, and our old friends. Turns out this was the project that I described in the chapter I wrote for Bob Reinhold's book.

The second win involved a project for the Boston Water and Sewer Commission (BWSC) and was important for two reasons. First, it had a large fee and covered two years. Second, we needed the work to keep our folks busy. It was work to eliminate combined sewer overflows in Dorchester and South Boston. As I mentioned earlier, this was a problem I had much experience with. Although I wasn't in agreement with the preferred solution picked by the Commission, sewer separation, I was more than happy to be paid to implement it.

We had a good team: me as project principal, Al Carrier as project manager, and Ding Fang as project engineer. I knew John Sullivan, the Commission's chief engineer loved our people, but our main (only) competition was Fay Spoffort & Thorndike (FST). Their team, led by their president, Emeil Hamway, was also well-known by John and had lots of experience in sewer separation projects. Emeil thought he had John in his pocket, and was arrogant and overly confident about winning. We played it low key and straight, and put on a flawless interview, with videos and bells and whistles. We came from behind and won with the interview. It was especially nice to beat Emeil, and on a practical level it gave much needed work to Al for the next two years.

4C. DARK DAYS.

There were two especially dark days at Rizzo Associates. The first came at a rest stop on the Mass Pike. We had a contract with the Turn Pike Authority to assess hazardous waste contamination at all of their rest stops. Of most concern were potential leaks from the tanks that held gas and diesel fuel. To test this, we drilled observation wells around the tanks and took soil and groundwater samples. One of the wells was between a rest area building and the overnight parking area for tractor trailers. At dusk, late one winter afternoon, while two of our guys were taking samples, a rig was backing out of the parking area and ran over one of the guys, killing him. This was before big rigs were required to have back-up bells, which might have saved him.

It was a terrible thing and Bill, God bless him, called the parents to alert them that their son had been seriously injured, met them at the hospital where he died and afterward, went to their home to return his personal effects. He said it was the most difficult thing he ever did, but the right thing. The other young man was shaken and was helped with counseling, but after some time left the firm. OSHA investigated and found that safety procedures were followed. We were fined a small amount, but the terrible feeling in the pit of our stomachs lasted.

The second time was not job related. We had hired a lead wetland biologist from the University of Rhode Island (URI), Mark Braiewa. He was terrific technically and could write. He and I were at a conservation commission hearing in Oxford, MA, about a strip mall one of our clients wanted to build. It was a wooded site next to a pond, and we were clear-cutting some trees and filling a small area of wetland. All in all, this was the type of project an environmentalist would hate. We were ready for a fight after Mark and I made presentations to the Commission. When we opened it for questions, no one had any, until the chairwoman said she had one. Mark and I braced ourselves as the woman asked if there would be a shoe store in the mall. I took a deep breath and said we didn't know but would surely take the request back to the developer. We got our permit. Mark and I didn't look at each other until we were in the car and then broke into laughter. Shoes were apparently more important than salamanders, at least in Oxford.

Mark and I went to many similar meetings and became good friends, although they didn't all end with the success we had in Oxford. After some time, we needed another good wetland scientist, and Mark had an idea, his wife, Amy. Amy and Mark had met in the master's program at URI and Amy was as good, or better, then Mark. That was saying a lot, and of course, Amy's qualifications were not the issue. The HR folks warned against having a husband and wife working together, and they were right. I agreed with HR, but Mark was persuasive, and we soon hired Amy. One of the best people we ever hired, but sadly we didn't have much time to test the husband/wife theory.

Mark didn't like doctors, and this turned out to be a bad strategy. He seriously cut his hand and didn't get the proper treatment. The wound got infected, and he still refused to go to the doctor, until he passed out one Saturday night and Amy called the ambulance. The next morning, Super Bowl Sunday, his best friend Andy called me at 7am and said Mark was in the hospital. By the time I got there, less than an hour later, he was dead. The infection became septic and shut down all his organs. The doctors said they tried everything. It was the saddest thing imaginable, as his mother arrived from Connecticut shortly after I did, to the same unexpected very, very bad news. It was just horrible. Bill and I spoke at his funeral, and I mustered all my strength to hold myself together. Amy remained a stellar colleague and eventually re-married.

Before we go on, a word or two about our first class capabilities in wetland science and permitting. In addition to Amy and Mark, who could do it all, we had Bernie Raftery, also from URI, who was unmatched in the field. Also, to perform his civic duty, he was on the Conservation Commission in Natick, our home town. Bernie was/is a big guy who normally got his way either on the Conservation Commission or before one. Bernie was what you called a low maintenance employee. He never complained, but rather told you 'don't worry, I have it under control', and he always did. Outside of work Bernie fishes for bass competitively and has taught me everything I know about fishing properly.

The best part about Amy, Mark and Bernie, they knew who was paying their salaries. For private clients it was usually a developer who wanted to build something. Our job wasn't to stop them, but to guide them in how to build it in the most sensible environmental way, following local, state and federal rules and regulations. If you want to protect every blade of grass, you shouldn't be in the consulting business. You should work for the Sierra Club.

Wetlands, which are critical environmental resources, cover a large part of Massachusetts and most large projects had to deal with them in some way. The State had a three tiered approach to wetlands. First you **avoid** them, second, if you can't avoid them, you **minimize** the impacts, and third, you **mitigate** the impacts that will occur.

An example, to illustrate the role of the consultant, might be helpful. We had a project to re-develop a large industrial area with 5 million square feet of a mixed-use, transit oriented development. In short, a 'good' project. When the site was originally developed for warehouse use, there were drainage ditches built to move water around the many large buildings and parking areas. Over the years the ditches had been neglected and wetland vegetation, mostly cattails, had grown as high as 10 feet. One of our wetland scientists had surveyed the site and wrote a memo essentially saying that the development potential was greatly reduced due to the cattails in the drainage ditches. They were a wetland species, protected by state regulation. Fortunately, he sent the memo to me to review, before sending it to the client.

After I blew my stack (you could hear me at the other end of the building), I asked for a second opinion. The Massachusetts Wetland Protection Act, one of the strictest in the country, didn't, at the time, make a distinction between natural and manmade wetlands, so our manmade cattails (of low environmental value) were under their jurisdiction. However, more seasoned folks pointed out that there should be no restrictions in maintaining the ditches as they were originally designed to move water.

The plan was to go to the Conservation Commission to get their approval to do the maintenance, since there was a legitimate conflict between protecting the cattails and preserving the drainage function. So I explained the situation to the developer, who had control of the entire site. Apparently he heard everything, except the part about getting the OK from the Commission. Before we knew it, he hired a landscape company to clean out the ditches, which they did in short order. Hence, the cattails went away and so did the conflict. The town never questioned what was done and issued a permit for the project. They even hired an outside wetland expert to give them advice. You can debate the developer's approach, but I never lost any sleep over sacrificing cattails, for a billion dollar redevelopment of an industrial site. Since that time, the Wetland Protection Act has been amended to specially allow what was done.

In addition to the two specific sad events mentioned above, there were a few times when we had layoffs, or had to let a specific person go.

Once the decision was made, Bill would delegate the implementation to either Rick or me, depending on the circumstance. Who likes to give bad news? It was normally me, because Rick's business area, hazardous waste, was strong and less impacted by economic ups and downs. I hated this because, although it was terrible for the people involved, I considered the layoffs my fault. After all, if I had brought in more work, it could have been avoided. Growth was hard when the economy was in the tank, but growth is what we strived to achieve.

Personally, I had a wonderful career, with many more ups than downs, but there were three occasions when I had a major crisis of confidence. Each had to do with a job I was responsible for completing successfully, on time and on budget. In the first two, one at M&E and one at Rizzo, I was behind schedule, over budget, and technically adrift. The easy answer was to just walk out and quit, and I considered this possibility. The first time at M&E, I froze up one morning and couldn't do anything. I took an early lunch break and vividly remember walking across Boston's City Hall Plaza thinking my life was imploding. Somehow, I ate lunch, went back to work, and got things back on track.

The second time was a similar situation at Rizzo, when I left work early one afternoon and went to the movies. Again, I thought my world was ending. I remember seeing "The Hunt for Red October" after reading the Tom Clancy book. It helped clear my mind, if only for two hours, and got me back on track. The final crisis came on a very large, mixed use development project. Simply put, I missed advising the developer about several significant problems on the site, related to wetland and wastewater issues. These could have derailed the project, and I lost much sleep over them for several weeks, until I navigated each problem successfully.

In all three of these cases I thought my failures were going to come to light and destroy my career. It was terrifying, but fortunately the lesson is, as Yogi said: "It ain't over until it's over," so keep working hard. The second lesson, which I never learned properly, was not to hide your problems and ask for help. Keep these lessons in mind, they are both important.

4D. GROWTH.

There are two basic ways to grow, one is organically and the second is through acquisition. Growing organically is preferred, mainly because it's less expensive. We did a good job locally by getting new work and hiring more staff. We also tried to expand geographically by hiring a key person in a distant location, New York City. We did this by hiring Morris Trichon, who was a PhD with good contacts in the city. He was a good guy, but we hired him expecting he would bring in new work. It's a chicken and egg thing, and he didn't bring in the necessary work, so we had to let him go, a painful event.

The quickest way to growth geographically is to purchase a firm in your area of interest. Unfortunately, we never had the cash to do this and as a result, never meaningfully expanded beyond Massachusetts.

Locally, Rizzo did purchase Sam Park Associates in the late 1990s. Sam Park was a transportation engineer, one of the best in the business. Sam wasn't content to be the best traffic engineer in the state, he was a real entrepreneur, salesman, and risk taker. He wanted to play on a bigger stage, as the owner/developer, not the hired help, with the goal of making much more money.

He had a small company of about 8 to 10 people located in Boston, and Bill was smitten by Sam's type A personality. The bad news: all but two people left within a year, but the good news was that the two people who stayed were Sam and Rick Bryant, the second-best traffic engineer in the state. Rick was an MIT boy and was terrific. He headed up our traffic work for a number of years, until he moved to Vermont and worked remotely, part time.

Sam was another story. He was brilliant, and like Barry Pell before him, could solve a traffic problem with a map, tracing paper, and a magic marker in 10 minutes. It would have taken a normal engineer 3 months. One afternoon, in about an hour, Sam did his magic for the state highway folks when they were beginning to consider eliminating the rotary at the end of Route 3, just before the Sagamore Bridge and Cape Cod. Many years and much money later, his original magic marker plan was built, essentially as he originally conceived it.

Sam's mind was always way ahead of the staff and that caused much confusion, and many mid-course corrections, as any project progressed. He was also a tad arrogant, but clients, especially developer clients, loved his aggressive ways. More to the point, Robert Kraft, owner of the Patriots, loved Sam, and as a result Rizzo Associates became the engineers of record for the permitting, traffic, and civil engineering associated with the new Patriots stadium. This was a big deal for us, and it alone almost justified the purchase of his company. It also might have helped that Sam's next door neighbor was Andy Wasynczuk, president of the Patriots at the time.

The stadium, in Foxboro, MA, took a detour while the Kraft's were trying to get the state to kick in more money for infrastructure improvements. The Kraft's threatened to move the stadium to Hartford, CT. To make things look real, we had meetings in Hartford and actually did some preliminary engineering. I believe, with no real proof, that this was all a charade, to get the attention of the folks on Beacon Hill. It worked, concessions were made (more public money), and the work in Foxboro began in earnest. No Massachusetts politician wanted to be accused of losing the Patriots to Connecticut!

This was a large prestigious project for us, as Sam designed all the traffic improvements (counter flow lanes, flyovers to avoid traffic signals, parking) and I took the lead in the permitting and infrastructure improvements. There were several flyovers (bridges) that need to be designed. At the time Bill was trying to get work in Poland, where they had well-qualified bridge engineers at very low labor rates, compared to the US. So we outsourced the bridge designs to Poland. They got it done, but we didn't save much money, partly because the Poles didn't understand the Mass Highway design criteria and the metric conversions, among other things. I believe if we had another five projects we would have worked out the kinks, but that never happened.

The second interesting design was the counter flow lanes on Route 1, which was Sam's idea. Before and after the game, one lane in the less traveled direction would be reversed to add a lane of incoming or outgoing traffic. This was done by protecting the reverse lane with Jersey barriers.

Also, we had a traffic engineer in a state police helicopter monitoring traffic and giving instructions to cops on the ground. The idea was to remove bottlenecks in real time. It worked quite well, helped by a new access road out the back of the site, taking traffic away from Route 1 in both directions.

On the environmental side, there were two improvements that were of major significance. The first was moving the Neponset River, which ran through the site. This was the headwaters of the river, so it was really a big brook, about 10-12 feet wide, running through the horse track on the site. The river ran under the track in culverts in two places and was an open channel in the infield. The whole track was to be demolished to make way for a large parking area adjacent to the stadium. If the river was to remain in its current location, it would run right through the center of the parking area and much of it would be an open channel, not in the Kraft's best interests.

We came up with a plan to divert the river around the proposed parking area, away from the stadium, and keep it all an open channel. The Kraft's liked this, but, to the environmental community, moving a river is like stealing your firstborn. Amy led the permitting effort and got it done in record time for several reasons. First, she was very good, but also the Kraft's put the pressure on at the State House. The key positive, which carried the day, was getting rid of the culverts and opening up much more of the river to daylight. It really was a good solution for everyone and today you can see the new, open Neponset River along the railroad tracks on the eastern edge of the stadium site.

The second interesting issue was water and wastewater management. When the original stadium opened in 1960, and everyone went to take a leak at halftime, the municipal water system couldn't supply the needed water for all the flushes, and the system lost pressure. This was a big fuck-up and no one wanted it to happen again. The solution was for the Town of Foxboro to build a new elevated water tank across Route 1 from the stadium, but even more inventive was to treat wastewater from the stadium, store it in a second elevated tank on site, and reuse it to flush toilets and urinals. This was a great idea, but the treatment plant, storage tank, and the reuse water piping cost $1 million more than just using potable water. The Kraft's were not interested in spending this extra money, so went to the town, twisted their arm, and got them to pay the difference.

When all was said and done, the Kraft's went public with their great environmental stewardship regarding the opening up of the Neponset River and the water/wastewater reuse system and, of course, so did we. What the Kraft's really cared about was getting a clear space for their parking behind the stadium and the reimbursement from Foxboro for the enhanced gray water system. You could say I should get over my cynicism, because it was a great project, and it was.

The real disappointment came after the stadium was completed on time, on budget, and to rave reviews. Because the project was so successful, we wanted to submit it for the annual award given by the American Council of Engineering Companies (ACEC) of MA. This required submitting as a team, with the owner. I went to Andy Wasynczuk to get his OK, which I thought was a no-brainer. Unfortunately, he was quite uncomfortable explaining to me that the Kraft's were in the process of suing us, along with the architect (HOK) and contractor (Skanska).

It turned out that Kraft's brilliant lawyers purchased an insurance policy that protected the owner from cost overruns. This was reasonable, since at the time (1999-2000) the construction cost was about $400 million. So a cost over-run of only 5% would be $20 million. The policy was expensive, over $2 million, but because the project came in under budget, there was nothing to collect on the insurance policy. This should have been good news, but the Kraft's, especially son Jonathan, was not happy they spent over $2 million and had nothing to show for it. The solution was to sue the project team for negligence, and collect money from the insurance policy to cover the cost of the negligent acts, although there weren't any.

Clearly, no one was willing to admit negligence, when none occurred. To avoid paying lawyers to fight, we all settled. We paid about $100K, HOK paid more, and Skanska paid even more, leaving a very bad taste in our mouths. Fortunately, decent developers, Trump being another exception, don't treat their subcontractors so poorly.

The Polish connection was my first exposure to international work and how things can get complicated. It wouldn't be my last, but they were all learning experiences. Poland was really Bill's initiative from beginning to end. As noted above, we had tried to expand geographically in New England and NY, with no success, so Bill got the idea to go international. Why Poland? Bill hired experts for Tufts Fletcher School and Poland was selected because it was pro-American, had massive transportation needs, and after the Soviets left, was a functioning free-market democracy. That's why.

Bill knew Thalia Schlesinger, who was Paul Tsongas's twin sister and well-connected in DC. Bill hired her to help get federal work, which didn't materialize.

Thalia was a great lady and knew everyone in DC. She made some Polish connections for Bill, and he was off to Warsaw. Bill did it all. Found an agent, opened a bank account, and started to develop business leads. Since 1989-90, when Poland had broken away from Russia, they were tilting westward toward NATO in a big way. Under Russia everything was state run, "thank you Commissar," but now everything was being privatized. For example, there were 8-10 district highway offices around the country and all were government run. Now, with privatization, they had essentially become their own consulting firms, each run independently. It was the one in Gdansk, Transpojekt, that designed the Patriot Stadium bridges.

After some false starts, Bill got us on a design-build team with a huge German contractor, Strabag, for about 50 miles of a limited access toll road, east and west of the city of Poznan. Poznan is about halfway between Warsaw and the German border. The master plan was to have this road, similar in design to the Mass Turnpike, run east-west from the Russian to the German border. Now, you might ask why, after being invaded many times by Germany and Russia over the centuries, would the Poles want to make the next time any easier for the tanks? Go figure, but we had a good laugh about it with our Polish friends. It was also interesting that when we asked them who they feared the most, Russians or Germans, there was never any hesitation. It was always the Russians. You can't trust the Russians. Somehow, they knew Putin was coming.

Now, this design build highway project was very large, and we were only a part of the design team, but it was going to put us on the map. As in most of these jobs, the contractor was the lead and called the shots. In this case Strabag was successful, and our team's proposal won the day. This was truly cause for celebration but we didn't know many details, as the contractor kept his financial bid close to the vest.

Apparently, part of the reason our team was successful was an aggressive (low) bid. It didn't hurt that through Bill's connection with Gerry Kavanaugh, Senator Kennedy's Chief of Staff, the senator call the Polish Minister of Transportation to put in a good word for our team. Since we were the only American firm in the competition, this was not a conflict of interest for us or the Senator.

When Strabag started to negotiate our fee, they played hard ball. This we could have handled, but the killer was they wanted us to assume too much liability, in case something went wrong. We were being killed on both ends with a small fee and large exposure. It was a deal you couldn't make, so Bill wisely walked away. It was a sad day. Unfortunately, this was a play that repeated itself, as you shall see.

At about this time, Tetra Tech was getting interested in Rizzo Associates and the Polish highway project was the kind of deal that would have given them heart failure. It might have even killed the purchase of Rizzo Associates. But, to pull a rabbit out of the hat, Bill suggested that Tetra Tech buy the engineering firm in Gdansk that had done the bridge design for the Patriots Stadium. The price was cheap, less than 1 million dollars for about 30 well-qualified engineers, and I thought it was a brilliant move. Tetra Tech didn't agree, and I think a great opportunity was missed.

The good news, for me personally, was a trip to Poland. As I mentioned, I was very involved with WEF and had been chairman of several national committees and task forces. An opportunity came up for two people to go to Poland on a technology transfer trip, funded by WEF. The topic of interest was environmental highway design, including drainage and erosion control. This was right in my wheelhouse and, best of all, it was to Poland. Given my WEF connections and qualifications, I was picked to go and asked to find someone to help me.

I found a highway guy from the NY DOT in Albany, and we planned our trip, which was to be one week, with 3 days of training in Krakow. Given Bill's efforts in Poland, I planned to meet him in Warsaw after Krakow, for a second week, and work on the design build highway proposal. It was a great plan. We had about 12 to 14 hours of training to prepare for the Polish transportation folks, and since my NY friend didn't have much to offer, I ended up doing most of the training. This all needed to be done in a classroom with a translator. I sent the written material and PowerPoint presentations ahead of time so they could be translated and copied.

The training turned out to be successful once I got used to working with a translator, and the rest of the time we enjoyed being tourists. It was May and we arrived for the first nice warm days after a cold winter. The square in the middle of Krakow could fit two football fields, with a big covered market in the middle and churches, hotels, retail, restaurants, and bars around the edges. We spent a lot of time sitting outside at bars watching the world go by. And what a world it was, since the oldest university in Europe was in Krakow and the co-eds were leaving their winter clothes in the dorms and showing off their long, lovely legs. It truly inspired our lectures.

When you went into a bar or restaurant, and we went in many, there were always two or maybe three pictures on the wall. The two were the Pope, John Paul II, a local boy, and JFK, another good Catholic. Sometimes there was a third, Frank Sinatra. Keep in mind Poland was about 95% Catholic, and the local priests had more clout than the major.

We took a field trip up into the mountains to see how a roadway was built along a small river, and the scenery was beautiful. One day, we were to be taken to Auschwitz, but I declined. It was so beautiful in Krakow, I didn't want to be overwhelmed with sorrow. The city of Krakow was spared destruction during the war, and the architecture was breathtaking.

Once in Warsaw, you could see how different it was from Krakow. It was destroyed during the war and the communists had rebuilt the city with block after block of the most uninteresting buildings you could imagine. An architect's nightmare, all in a drab gray color. There was one square that was surrounded with reproductions of buildings that had been there before the bombing. They were colorful and beautiful.

Bill and I stayed at the wonderful Bristol hotel, which was mostly spared during the war and refurbished. The first night we were sitting in the wood paneled bar waiting to have dinner with some connections Bill had made. In walked two of the most beautiful women we had ever seen, and they sat down next to us at the bar. It took me less time than Bill to realize these lovelies were hookers. At first, I think Bill thought they were just nice ladies looking to chat us up. We had a good laugh at dinner with our local friends, who confirmed our suspicions. In hindsight, if you were going to stray, they would have been on the top of your list.

We took a side trip to Poznan, to check out the roadway project and make some contacts. We stayed in the smallest hotel room ever, but it had everything you needed –just don't try to turn around with your toothbrush in your mouth. When we met our contacts, the ice-cold vodka flowed freely. When I got back to the small room one night, after many toasts and much laughter, the only place I could fall was on the bed. I slept well.

The communists' legacy was all around, but my favorite illustration was the church in the middle of the main square in Poznan, which was much smaller than Krakow. There was no upkeep of the church during the Cold War, or much else for that matter, but the new leaders in town were trying to catch up on all the deferred maintenance. They were sand blasting the church walls and were about half done when we were there. The half that wasn't done was black and none of the details were clear. The half that was done uncovered an architectural masterpiece that Michelangelo would have been proud of. It highlighted what a disaster communism was for the people of Poland, and elsewhere.

4E. HOME LIFE.

One of the attractive things about going to work for Bill was the commute. From our home in Wellesley Hills to the office in Natick was 15 minutes, and reliably so because it was against the commuter traffic.

I didn't want to leave Boston, but Wellesley wasn't exactly slumming it. Erin went through Wellesley public schools, got a terrific education, and made lifelong friends.

We had a small house among mini mansions, on ½ acre, with a wooded lot behind us. As someone who visited said, it was the perfect suburban, "Leave It to Beaver" neighborhood. I was appointed to the conservation commission and served for 7-8 years, chairman for the last few. Wellesley was a well-educated town of about 27,000 people. We had a very professional commission, with lawyers, biologists, developers, and me. No one could throw a fast ball by that commission. I made some good friends and did my civic duty.

Wellesley had a town newspaper, the Townsman. It came out once a week, chock-full of local news. Sometimes the conservation commission was in the news and from time to time, when I read the article, I wondered what meeting the reporter had attended. When they got it wrong and I was involved, I would be awake all night asking myself why?? It was clear I had a thin skin and was not cut out for politics. One time I was accused of a conflict of interest by one of the engineers representing his client before the commission. It was total bullshit, but it's hard to fight fake news once in print. Fortunately, things were mostly quiet and professional.

Cynthia continued to work when we moved to Wellesley and, until Erin got a spot in the Wellesley College daycare, we all commuted to Boston every day. Soon one car wasn't enough, so we got a second. Cynthia worked for New England Nuclear (NEN) in the South End and loved her job. The company was driven by PhD chemists, who developed a technique to "tag" compounds, mostly drugs, with radioactive isotopes. When the drugs were taken during testing, the medical folks could follow where they went in the body by following the radioactive material. This was a key part of the medical research associated with developing new drugs, among other uses. The radiation was of such low concentrations that it wasn't harmful. Cynthia organized marketing materials for grant applications, mostly to the federal government, and did much of the writing/editing for the brilliant scientists. It was work similar to what Erin has done in the medical industry, where there is a need to explain complicated things in clear, understandable language.

The company was very successful, so it became an acquisition target and was purchased by a large pharmaceutical company. The whole operation was to be moved to Billerica, MA, making for an unpleasant commute, so Cynthia left. She looked for another job nearby, but none came along so she stayed at home taking care of Erin full time.

We took all our vacations in Maine, which is what we both wanted to do. We would also travel to see my sister, Leslie, and her family in Swarthmore, PA, or Cynthia's sister, Emily, and her family in Cherry Hill, NJ. Other than that, we stayed home. I was 100% committed to work. There were many stressful days, but I loved what I was doing and worked nights and weekends to get it right. Cynthia was also quite content with this arrangement, as she was not a joiner. Although not an outgoing person, she was an extraordinary judge of character and could spot a phony across the room. Her analysis of people after she had met them, even for a short time, was uncanny, and very helpful to me, since most of the people she met were colleagues or clients.

I spent a lot of time with Erin on the weekends before she went to high school. I frequently went to work on Saturday and Erin would often come along. She would do schoolwork or play in my office, but her favorite thing was to go into the big conference room and pretend to teach an imaginary class. She would talk to her "students" and use magic markers on the white board to make her point. One time she didn't erase her markings and there was some discussion Monday morning about what was going on over the weekend.

Neither Cynthia nor I were church goers, quite the opposite, but church was a big social deal in Wellesley. We told Erin if her friends were going to Sunday school and she was interested, we would take her. She wasn't interested, so I decided to have Sunday school at home. I considered these to be history/geography lessons, not religious ones. We covered the standard Bible stories, some exposure to other religions, and looked at lots of maps, my personal favorite. We figured Erin could make her own decisions about religion later in life. One benefit, which we still laugh about, had to do with a geography lesson. I wanted Erin to understand the globe, so we talked about continents, mountain ranges, and major rivers, using some great maps. Along with the Rocky Mountains and the Himalayas, I pointed out the Ural Mountains in Russia. Erin was in 5th or 6th grade at the time, and it just so happened that her teacher was also doing a geography lesson and asked the class if anyone knew the mountain range that separated Europe and Asia, not expecting an answer. Erin, without missing a beat, confidently said the Ural Mountains and gained many points with the astonished teacher. It was also a lesson on why lessons are important.

About 8 years after my experience with testicular cancer, Cynthia was diagnosed with ovarian cancer. This was only a few months after her annual physical, which uncovered nothing. Unlike my cancer, hers was not so successfully treated. After a round of chemo, things looked good for a few months and then the cancer came back. This was definitely not good and we were told to prepare for the worst, which was surely to come. It didn't take long before the pain was too bad to stay at home and she went into the Youville Hospital in Cambridge for pain control, as the end came. Cynthia didn't want to see people in her final state and especially wanted to shield Erin from any unpleasantness. This of course was not possible, but noble to think of others at such a time. Her strength and calmness, facing the end, were remarkable, especially since she was only 48 years old.

Erin was a freshman in high school, so with the help of family and local friends, we made it through her high school years. So many of Erin's friends and their families looked after her and gave meaning to the notion that "it takes a village." I took the approach of a "stiff upper lip" and thought keeping very busy would serve us well. It was my mistake. I should have given Erin and I more quiet time to grieve properly. One of Erin's friend's fathers had died, so we went to a group with families that had lost a parent, but it wasn't helpful, at least for me.

I spent Erin's high school years buried in work and the two of us settled into a routine. I was home for dinner, except those nights I had meetings, and did a lot of simple cooking. Erin excelled in school, was self-sufficient, especially after she could drive, and was a great, responsible kid. I couldn't have asked for an easier teenager, given the risks at that age from drugs and alcohol, which were huge problems at Wellesley High School. The senior prom was canceled Erin's junior year because of alcohol problems on a class trip and at a number of pre-prom parties which were unsupervised by adults.

We had a developer client, Steve Cohen, who was a great guy, and had built a condominium project on Nantucket. He had kept one of the units and it was so nice, it was featured in House Beautiful. There was a drainage problem on the property and one day he was explaining it to me. During the discussion he offered to fly me out and stay in his unit, in exchange for me looking at the problem, and Erin could come along. So we did and had a great weekend, in the off season. Nantucket is a beautiful place year-round. On the way back, the pilot let Erin sit in the co-pilot seat, and it was quite an experience for her.

The summer before Erin's senior year, we took a trip to Paris and Rome to see the sights. We stayed at the Ritz in Paris and Erin was impressed by the fact that Princess Diana had also stayed there. She was a great admirer of Diana and was crushed by her untimely death. Rome was fine, but hot, and we were ready to go home. Her senior year we visited Bates, where Cynthia went and graduated Phi Bata Kappa. The dean of admissions, who interviewed Erin, had gone to Wellesley HS, so all this conspired for an easy early acceptance. Leslie had taken Erin to see colleges in Pennsylvania, including Lafayette, but she wasn't impressed. I was happy she was going to Bates, especially given the short 2 ½ hour ride.

Early on, starting at M&E, Robb, Tom Lynch, and I spent Father's Day weekend in Maine fishing. Later Frank Murphy made a foursome, and we added golf to the mix. We did this for almost 40 years, until Tom, who moved to Florida, developed some health issues. The reason we picked Father's Day was because the fishing licenses were free in Maine that weekend. For several years, after Robb bought a farm in Sandwich, NH, we went there and fished in Squam Lake.

Frank would bring four huge steaks from Savenor's on Charles Street in Beacon Hill and grill them medium rare. We would all claim we couldn't eat the whole thing and then proceed to do just that. After, it was a fire on the dock, much wine, star gazing, and good conversation. We looked forward to this time together.

Tom was a lifelong friend who died a few years ago. He could be difficult and often said things that, although perhaps true, people could take the wrong way and did. In his own way he cared very much for his friends. When I was 39 and going through chemotherapy, he along with my friend Jim Schwarz were the two people who called me every week to see how I was doing and offer support. A window into what people are made of, and something you never forget.

The Maine cabin has been the one constant in my life and Erin's. Cynthia and I bought the cabin when she was pregnant with Erin in September 1980. That following spring I was opening the cabin with Peter Swartz, who also had a cabin on Little Sebago Lake. We were in his pickup truck passing a red wooden rowboat for sale on someone's front yard. It was only 40 dollars, without oars, so I asked the guy if it would float. He looked at me and said in that distinctive Maine accent, "For 40 bucks what do you expect?" I bought the boat, and we hauled it back to the cabin in Peter's truck. Later, I bought a set of oars, which cost more than the boat. It did leak, but not too badly and we had it for years.

I went fishing after dinner most nights, and Erin often came along. We didn't go far with just oars and caught mostly sunfish. It was, and still is, the sunsets that are the real treat. I smoked a pipe and would tell Erin stories about Matilda, a mythical Australian girl about Erin's age. At one point I made a wooden scarecrow of Matilda and we bought clothes at the local antique barn to dress her. I also carved a wooden pipe for Erin to amuse her while I was telling stories. It was a quiet, wonderful time, with no stress, even if we didn't catch many fish.

4F. THE ENDING AND NEW BEGINNING.

Rizzo Associates lasted about 18 years, and it was a great run with terrific people, but we didn't make much money and the relationship between Bill and Jim was not going well at the end. We also had Sam Park on board, with his endless ideas. Some were indeed brilliant, but it was hard to separate the good from the crazy, especially with Sam's spin applied.

In the end I think Bill was tired of Jim's constant pressure. We had become an ESOP, which Jim didn't like, and Bill just had had enough. He wanted to have an "inside sale" of the company. His preference was to sell Jim's shares to some combination of Rick, Liz, Sam, and me and to have one of us emerge as the new president. I also believe this was a non-starter for Rick, who wasn't in the position to buy it all and surely didn't want to be financially tangled with Liz and Sam. Liz and Sam knew this and came to me with a proposal for the three of us to buy Bill and Jim out. Like Rick, I had financial limitations and also thought that being partners with Liz and Sam was crazy. From day one, there would have been constant conflict about the direction of the company, given that our styles and priorities were vastly different.

The other problem was Jim's take on any transition. Bill was hoping that with an inside sale Jim would agree to a discount in the price. This he was not willing to do, and wanted full market value for his stock.

So Bill was left with the option to sell the company to an outside entity. He hired a mergers and acquisition guy, Paul Zofnass, one of the best in the business, and the bidding began. Bill picked the right time, a stroke of good luck, and we all benefited financially from the sale to Tetra Tech, headquartered in Pasadena, CA. Tetra Tech was founded and run by Li-San Hwang, a brilliant engineer, with a PhD from Cal Tech, and a natural businessman and entrepreneur. He was financially driven and contradicted the notion that technical folks aren't good business people.

Li-San had built a 2 billion dollar publicly traded company, based on acquisitions, and we were in his sights. He paid too much, in part because another bigger deal had just fallen through and he needed our revenue to make his next quarterly financials look good. He was an extraordinary guy, partly illustrated by the fact that Tetra Tech's stock price increased through the 1990s faster than Microsoft. Because Rizzo Associates was an ESOP company, everyone got a check equivalent to their stock value, when the company was sold. Those who complained when we became an

ESOP (they wanted cash bonuses, instead of stock) had smiles on their faces, as we all did.

Bill and Jim made out the best of all and Jim illustrated his knack for squeezing as much cash out of a stone as possible. His agreement with Bill stated that he would be on the board of the company for life. So legally he was entitled to be compensated for not having that opportunity, to the tune of several hundred thousand dollars. This was a big shock to us, and seemed unfair, but his legal argument was sound and there was nothing we could do.

Despite the financial good fortune, there was concern among the staff, especially senior staff, that "big brother" from Pasadena, CA was coming to micromanage, put profits above everything else, and change our culture. These were legitimate concerns, given Tetra Tech's reputation for financial discipline and the horror stories you hear about other acquisitions. Some of this happened and some didn't, but from where I sat, it was a terrific deal for both sides and still is.

CHAPTER 5

TETRA TECH, 1999-2008

As I alluded to above, you can't talk about Tetra Tech, before say 2010, without understanding Li-San. He was born in China and came to Michigan to get a PhD in fluid mechanics. His advisor apparently realized that Li-San knew more than he did and sent him to Cal Tech to be properly challenged academically. Early in his career, during the Cold War, when he was doing technical stuff, he helped design the shape of nuclear subs for the Navy, so they would be quiet moving underwater, and not be detected by the Russians. He also helped design the computer program for the first tsunami warning system in the Pacific Ocean. Unfortunately, it was not fully deployed when the big one hit in 2004 in the Indian Ocean, with waves higher than 150 feet. Clearly, Li-San was a smart guy, but his true genius was building a company and managing the details, and I mean details.

Li-San was running a $2-billion company, yet you would get a call from him about an invoice that hadn't been paid for over 90 days. He wanted to know why, and you'd better have an answer. The anxiety of getting such a call, and I did, kept you up-to-date on all your accounts receivable. It was one of his many management tools that kept you on your toes.

Li-San kept track of everything , and led by example. He was worth more than $100 million, but would fly coach. We had annual meetings at nice places, on the west coast. Several times they were at the Ritz resort in Henderson NV, just out side of Las Vegas. The years we did very well financially I had our contingent (5-6 people) fly business class, rent fancy cars and generally enjoy the fruits of our labor. One time a few of us were flying back from the annual meeting and just so happened that our flight was leaving the same time as Li-San's flight back to Pasadena, at adjoining gates. As the two groups were chatting I realized that if our flight called for boarding business class passengers, Li-San would be on to us. So I quietly told our group to wait and board with the coach folks. A small point, but it illustrates how Li-San's philosophy became yours, even if you cheated on occasion.

Tetra Tech started and essentially remained a beltway bandit focusing on all aspects of water, using high end computer models. They also did a lot of hazardous waste clean-up work, much of it for the military. In the late 1990s Li San branched out in two directions, both related to infrastructure. First by buying firms that did water and wastewater work, mostly for municipalities. That included us. We were one of six or seven regional mid-size design firms scattered around the country, in our case New England. Other firms were in Florida, Kentucky, Colorado, Washington, and California. None of us did any serious federal work. The second was in fiber optic cable construction during the dot-com boom. These were actually construction firms building out the fiber optic cable network throughout the country. Laying cable is not unlike building water mains, sewers, and storm drains.

By the time we were bought in 1999, Tetra Tech had 30 operating units in three basic areas: the original water work, mostly for the U.S. government, water and wastewater design for local clients, and fiber optic construction nationwide. Most of the acquisitions had annual revenues in the $20-50 million range, with a few over 100 million. Li-San was dead set against overseas work, despite his China connection, or maybe because of it. His only exception was overseas U.S. government-funded work, for organizations like the military and USAID. Li-San's concern was mostly about corruption and getting paid, and he was right. He knew he would get paid from the U.S. government. I wonder what he would say about Trump's gutting of USAID, which was and is, the most cost-effective tool for projecting America's interests abroad. It was far from perfect, but better than another Abrams tank at $10 million, or another destroyer at $2 billion.

Safe to say that Li-San was financially driven. I suspect that might have been because he was so gifted technically, and he assumed technical quality was a given. He also knew that technical people generally hated the nitty gritty of financials, especially collecting money in a timely fashion. He didn't like working for state transportation agencies, because their salary and overhead caps were crushing. Of course, Bill's main connections were in transportation, so he nicely explained that the key to doing transportation work was keeping your overhead low by diversifying and filling niche areas, such as hazardous waste. Bill applied some spin to his arguments, it worked, and the deal went through.

Li-San's business philosophy, at least regarding acquisitions, was to let each run their business, but keep a close eye on financial results. Every year, you came up with a financial plan for the next year and were expected to meet it. You were always pushed to make the plan a bit aggressive, but if you met the numbers, you were free to run your show. There were quarterly investor calls, like every publicly traded Wall Street company, and Li-San was brilliant at managing these calls. As he explained to us, once the investors accepted the overall financial projections, they wanted the roll-up of all 30 operating units to meet them. Falling short in revenue or profit projections was, of course, a problem for the stock price. Doing much better was also a problem, despite what you may think. To the investors, it looked like you either didn't understand your business or were sandbagging the original projections, so they punished you. In short, you were expected to hit the numbers. Li-San was a master at this by managing the operating units and adding more all the time.

Each unit was a wholly owned subsidiary, with separate income statements and balance sheets. This made tracking financials easy from a corporate standpoint. Internally, Li-San had a simple formula that boiled down all the financial "bean counting" into something we were all keenly interested in, namely our bonus calculation. It was based on the fact that Tetra Tech corporate got the first 6% of the profits. For profits over 6%, 75% went to corporate and the remaining 25% went to the operating unit to form the bonus pool. This pool was adjusted based on your performance in collecting money. The goal was to collect money in 70 days. If it took longer, your bonus pool was reduced based on the cost of borrowing money to cover ongoing costs.

This bonus formula was a terrific system that was clear and required no negotiations. For example, if your revenue was $10 million, a 6% profit would be $600,000. Let's say you made 10% profit, your bonus would be 25% of the additional $400,000, or $100,000. This would be adjusted based on your collection history, which usually was not a major adjustment. The bonus was given to the president of the operating unit, who had full discretion on how it was distributed. Another smart move by Li-San. On top of the cash bonus, stock options were given at the discretion of corporate.

The bonus process was the end result of a year's hard work. So the question was, how do you manage a business throughout the year to get the best result? Here Li-San also had the answer. He argued that the most important metric to monitor, on a weekly basis, was utilization. Companywide, if more then 65% of the total hours are being billed to a client, you are in a sweet spot. The non-billable hours included holidays, business development and the like. The most meaningful way to make the calculation was to factor in salaries, so an hour of non-billable time by a VP is weighted more than an hour of a staff engineer.

There are many other metrics to monitor including accounts receivables, budget overruns, cash, and of course project profitability, but these are all backward looking, as is utilization. Beside utilization, my favorite metric was backlog, because it was forward looking. When you look backward you have reliable data, so long as you put good information into your system. Looking forward, with backlog, required a healthy amount of judgement. If you have a signed contract, your judgement is solid, as long as you know the work schedule. If you were submitting a proposal, you may estimate your chance of success at 50%. The point is, estimating backlog was more of an art than a science. You needed to know your business to perform a reality check on the numbers. Six months of reliable backlog was a good place to be.

In a company with a well-managed financial system, like Tetra Tech, there were many metrics that help you manage the business. The trick is to understand them and act accordingly. It's easy when the numbers are good and not so easy when the number are troubling. The most unpleasant actions that result from weak financials and/or low backlog are layoffs. It's human nature to avoid this as long as you can, but that is a mistake, as painful as it is. A more impactful decision could be to not work for a certain client or client type, mostly because they don't pay enough (profitability), or don't pay at all (receivables). Needless to say, running a business when things aren't going well financially is stressful.

We were purchased by Tetra Tech in 1999. Shortly after, the dot-com bubble burst and the fiber optic gold rush abruptly ended. Apparently, some hedge fund had offered Li-San over $1 billion for his fiber optic business before the crash. He refused, one of his few mistakes, because shortly after, the business was worth next to nothing. This brought the stock price crashing down and instead of retiring, Li- San stayed on for several more years to right the ship. This he did by getting out of the fiber business and focusing on our core businesses. The bright spot for us was the low price of the stock options we got during this period, which 10 years later were worth some real money.

As mentioned, joining Tetra Tech was a good thing. They did what they said they were going to do, namely stay out of our day-to-day business and watch the money. For example, we wanted to keep the Rizzo name for marketing purposes and to bolster staff morale. So we became Rizzo Associates, a Tetra Tech Company, when other parent companies would have immediately buried the Rizzo name. It may seem a small thing, but made a big difference, and Li-San understood this.

On the money side, Li-San immediately set up a meeting with another recently acquired company in their NY City offices. This was considered, by us, a kickoff meeting, with lots of hugs and kisses. Several of us went down to NYC and Bill made a wonderful PowerPoint presentation about how successful we were. Li-San was there with his CFO, Jim Jaska, and after the presentations, calmly thanked Bill in his thick Chinese accent, before cutting to the real reason for the meeting. He made clear that our financials were a long way from being acceptable. He especially focused on the amount of time it was taking us to collect money, about 130 days as compared to his goal of 70 days. He did not sugarcoat his concern and pointedly ask Bill to commit to an initial reduction to 90 days in a few months, with the longer-term goal of 70 days. Bill, a bit stunned by Li-San's tone, as we all were, really had no choice but to agree. Li-San could be charming, but not that day, as he was staking out the rules of the game. This was not a private conversation with Bill, we all heard the unvarnished message, both companies. After the meeting, which didn't go as we planned, Li-San bought us all an excellent meal at a fancy restaurant, with much whiskey and wine, but the conversation was subdued.

Bill decided to transition out of a leadership position before Tetra Tech, but I suspect this meeting accelerated the process. He favored Rick Hughto as his successor, especially since Rick's approach to financials and how to run a business were lined up well with Tetra Tech's. It was actually too good to be true. Rick was a western version of Li-San, but as Bill talked to several senior people, he realized that we would leave if Rick was named president. We all, except Liz and Sam, liked Rick, but had a hard time with his leadership style. He communicated mostly by writing and didn't promote the face-to-face, give-and-take that Bill, and the rest of us, enjoyed. It would have been a top down, as opposed to a collaborative, organization, as it was under Bill.

I would have been happy to leave and had told Bill years earlier that I would likely leave when Erin went to college. So it was time, until Bill had second thoughts about Rick taking over. As a result, he offered me the job, and it was hard to say no, so I didn't. Of course, the opposite happened, and Rick left, as did Sam. Both were not a surprise, and fortunately no one else left. I think I can take a little credit for that. I maintained a good relationship with Rick, who started his own business focusing on being an expert witness in hazardous waste cases. He knew the business inside and out, had excellent credentials (PhD), and the lawyers loved him. I offered him office space, which he could use at no cost, in anticipation that he would use us if he needed support (he had no employees) and give us leads if he had them. He mostly worked at home and we never got any big jobs from his leads, but we had a good relationship for many years. I did some expert testimony work that Rick referred to me, but I didn't like how nasty the lawyers could be, especially during depositions.

Tetra Tech could overrule almost any local decision, but rarely did, outside of financial decisions. Bill made a good case for me with Li-San and the deal was done. Bill was very gracious in organizing the PR for my transition to president and things went smoothly. He quickly backed away and let me run things. A good move for everyone. There were several issues to deal with that set the stage for the future. Not surprisingly, the important ones had to do with personnel. But first a word about organization and how to sell our services.

The two general ways to organize are around technical disciplines or around clients. The decision is often dictated by the talents of your senior people, and in my case, largely because of this, I chose an organization with a little bit of both approaches:

1. Hazardous waste, Ray Johnson
2. Private sector development, including traffic, Brian Sullivan/ Rick Bryant
3. Environmental permitting, including highway design, Ed Ionata/Chris Calnan
4. Federal programs, Rich Sperandio

I had Ray, Brian, Ed, and Rich as direct reports, and that was enough. These were the people, including Rick and Chris, who were going to make it happen.

Now, Bill had brought on some other senior people, all of whom I didn't think added value commensurate with their salary, especially given that we were now a company driven by the bottom line. Keep in mind that while I was president, I was always about 50% billable. Senior people clearly needed to justify their salary by either being billable or bringing in enough work to keep dozens of other people busy.

There were five areas where I had concerns. The first was in transportation. Bill had hired several senior people, part time, who had essentially retired from other companies that worked for MA DOT. They were all good folks and knew the public sector, but poor Chris Calnan didn't really know who was in charge. He got lots of advice, but nothing seemed to lead to a real job. These folks, I think there were three of them, were nicely phased out to pasture.

The second area was telecommunications. As mentioned, this was a big deal until the bubble burst in 2000. Bill had hired a woman with background in the business, although she was not a technical person. Since she couldn't do technical work, and was paid a lot, she needed to bring in enough new business to keep 5-10 people busy. Our main business was locating and permitting cell towers, but not enough of this work was happening. Her second in command, Mike Turcott, was a competent engineer, who in my opinion was just as good, maybe better, in business development. Her overhead cost needed to be cut out, and her second in command promoted to run the show. It was a no-brainer and done quickly, with no short term negative impact, except she sued us (me) for bias against women. We settled.

The third area was finance. Bill had hired a CFO at a high salary. This was OK, especially during negotiations with Tetra Tech, since the guy was smart and could talk a good game. The problem was he operated from 30,000 feet and didn't get his hands dirty. We were about to change our entire financial system to Tetra Tech's, and I needed someone who was willing to get in the trenches. Also, he was a terrible people person and did not respect the women working for him. I fired him and it was one of the best personal decisions I ever made. I hired a young woman, Susan Demarre, with an MBA from Wharton, who got in the trenches and was terrific.

Another note on financials. Before the Tetra Tech deal, we had decided to upgrade our financial accounting system and the CFO had picked the IT/software company we were going to use. When we started to talk to Tetra Tech we put the transition on hold. Good thing, because the IT company went bankrupt 6 months later. Tetra Tech was also in the process of bringing all their 30 operating units onto one financial platform. They were using Oracle and wanted one of the operating units to volunteer to be the first to convert. With Susan as our new CFO, we jumped at the chance. It was a great decision because we had unlimited support from corporate and Oracle in making the conversion, and Susan got to show her stuff. Everyone won.

The fourth area was the most difficult, since it dealt with Liz. As noted above we worked closely together for 15 years and were friends. Liz's husband, Chuck, was a great guy and they had two nice sons. One weekend, Chuck, the boys and I went fishing at our Maine cabin. Liz was also very fond of Erin, taking her to tea at the Ritz. You get the point. Unfortunately, after the sale to Tetra Tech we were in a different universe. Bill had been, for years, trying to get Liz more billable, as evident by making her head of the transportation group. It hadn't worked and there were no consequences.

When I became president, I told Liz things had to change and quickly. I could have said it was business not personal, but that's bullshit. At some level I thought Liz's philosophy was 180 dregs opposite of Tetra Tech's and that she was doomed to fail. Although I gave her an option to change, I really wanted her to leave, and my position likely came off that way. Liz apparently thought so too, didn't leave quietly, demanded a crazy severance package, and said some unkind things about me to her friends, including that I was out to get women, something she knew was not true. Liz also asked Bill to get involved and influence me, which he didn't do, a painful decision for him. In short, it was messy, but without a doubt the best thing for the company. It was a good business decision as we couldn't afford a $200,000 salary with little to show for it. Also, I knew that Tetra Tech would point this out sooner or later, and I wanted to get out ahead of them. I have no regrets for the outcome but do regret the clumsy way I got there.

The fifth area was much more positive, and also had to do with one of Bill's hires. Rich Sperandio had worked with Bill at M&E and I knew him as well. He was easily the best program manager I have ever worked with, and he had deep experience working with the federal government. Bill thought very highly of Rich, for obvious reasons, and hired him after years of trying. The good news was we had Rich, the bad news was we had no work for him and building a federal business takes time. Keep in mind that when he started, we were just Rizzo Associates, a mid-sized firm, working almost exclusively in Massachusetts. We had no track record with the federal government

In the beginning, the late 1990s, Rich helped me with the Patriots Stadium work, but he was not fully utilized. Rich was frustrated and wanted to pull his weight. Fortunately, another M&E alumnus, Ed Greco, was working at Perini, a large construction company, also located in Framingham, 10 minutes from our office. I knew Ed well, but he was tight with Rich and Bill. Ed was the lead business development person for Perini's international group, which mostly worked for the military overseas. After some false starts, Rich formed a three-way joint venture (JV) with Rizzo, M&E, and AECOM. Rich was the lead of the JV. Subsequently, AECOM purchased M&E and Tetra Tech purchased Rizzo. So the JV then consisted of two of the biggest consulting engineering firms in the country, supporting the contractor, Perini, in design-build projects for the military. The Perini team looked appealing to the military because of its size and experience. We could now play with the big boys.

This was in Rich's sweet spot, given his experience in managing the design part of design-build projects. He had worked for the Air Force (AF) on airfields in Saudi Arabia and on early warning radar sites along the Dew Line, while at M&E, in some cases with Ed. So Rich and Ed had a very strong relationship. They knew the AECOM folks, who were also old M&E guys, mostly guys.

A typical project is designed by the engineer/architect and put out to bid for contractors. The project owner has separate contracts with the engineer and the contractor, and typically the owner looks to the engineer as their representative during construction to "keep the contractor honest." In the design-build model the engineer and contractor work as a team with one contract with the owner. In this case the engineer is essentially working for the contractor and, unlike a standard project, doesn't need to complete a detailed design that would normally be required for bidding. The engineer only needs to advance the design as much as the contractor needs for a cost estimate, to win the job, and to build the project once selected. Many engineers have trouble with this approach because they want to design everything to the last detail. Good contractors don't need this detail and don't want to spend the money to get it. Keep in mind that in design-build projects, the contractor's bid includes the engineer's costs, it's not paid separately by the owner, like in design-bid-build projects.

Rich understood the design-build method and Perini loved him for it. The first big job we went after was a large radar site on the top of a mountain in Taiwan. The Air Force (AF) was the client, Raytheon was building the radar, and the Perini team was to design and build the facility to house the radar, including an access road up the mountain. Rich went for a site visit with Perini, and we prepared enough engineering detail for Perini to estimate a cost for their bid. We spent about $100K to help Perini prepare the bid, anticipating that we would recoup most of these costs if we won the job. $100K was a huge amount for us to spend on speculation, but we were playing in the Big Leagues now.

The good news is we won the project and couldn't believe our good fortune. Needless to say, Rich was ecstatic and largely responsible for us winning. Soon after, Perini received a draft contract from the AF. It included language that placed a lot of risk on Perini, if schedules and budgets were not met. It might also have had a liquidated damages clause, but I don't recall. In any event, the president of Perini, Bob Band, was a financial guy with a low tolerance for risk, especially considering this project was on a mountaintop halfway around the world. Perini also had a group that built casinos, and it just so happened that they had just been awarded a billion-dollar contract to build a casino in Las Vegas. Not surprisingly, Bob wouldn't sign the Taiwan contract without changes that the AF rejected out of hand. So he said, "fuck you" and walked away, knowing he had plenty of low-risk work in Las Vegas. We went from joy to despair and Rich was crushed. An ending not unlike Poland, but with a much better second act.

As it turned out this likely would have been the project from hell, but we didn't think so at the time. I took the set-back pretty well and told Rich not to despair because we had an excellent Perini-JV team and there were more opportunities to come. Little did I know that 9/11 was just around the corner, followed by endless wars in Afghanistan and Iraq. Whether you agree with them or not, we profited from the wars and the endless projects the U.S. funded in those countries. Most of the building projects were managed by the Corps of Engineers. The AF also had a lot of work, including early warning radar sites in England, Greenland, and Alaska, essentially the rebuilding of the DEW Line.

To say that Rich took advantage of all this work, both with Perini in Iraq and separately with the JV in Afghanistan and for the AF, would be an understatement. In short order he had 50 people working for him in a very profitable group, all Tetra Tech employees. AECOM had just as many. In Afghanistan the work was real nation building. At first it included building army battalion facilities for about 500 troops from scratch (roads, buildings, utilities). There was nothing useful in-country, it was bombed and bombed again, with piles of rubble everywhere, including unexploded ordinance. Everything had to be shipped in, over the Kiber Pass, and small enough to fit through tunnels in the mountains. Interesting logistics. Later the work included public buildings and schools, for boys and girls. Today much of this investment is in the hands of the Taliban, with very different ideas.

In Iraq, most of the work was in direct support of the army. The biggest project was designing covers over mess halls, PXs, and housing, to absorb the explosion of incoming rockets and protect the troops below. These were big structures given the size of the buildings below. We were paid for each cover on a lump sum basis, and after we did the first 5 or so, we got very efficient at the design process. As a result, our costs went down, but the lump sum price stayed the same for many more facilities. Given that our original price was conservative, after all we were in a combat zone doing something no one had done before, we started to make a significant profit on each cover. As the saying goes, big risk, big reward. We received an ACEC award for this project, the one we never got for the Patriots stadium.

Large war profits were not always the case for Tetra Tech. Since Tetra Tech traditionally didn't do much overseas work, one day the Board of Directors asked Dan Batrack for an up-date of international work, most of it for the military. So Dan asked each operating unit for a financial summary of their overseas work. Since there were only two units in this category, we each responded. The other unit was much larger than us and doing about 50 million dollars of work under a subcontract with another company, making only about 3% profit. We, on the other hand, were doing about 10-15 million dollars of work, but making more money (profit) on much higher margins. This was mostly because of our lump sum work. When Dan looked at each spreadsheet, he called me and said there must be some mistake. I assured him there were no mistakes and he happily told me to do more of the same.

It pains me to say that we profited from war, and other people's misery, but we did. But I am happy to say that we sent many folks to a war zone as part of these projects, all volunteers, and no one got hurt. There was one ambush where a Perini guy got a minor leg wound, but none of our work sites were ever attacked. Many companies would not go to a war zone, but we did take the risk and fortunately things turned out well for us.

When I took over as president, the economy was in the midst of the dot-com recession and things were tight for the first two years. As I mentioned, our telecom group was shortly dissolved, and we had to work hard to avoid layoffs. Rich was starting to build the international group, but it wasn't there yet. I didn't need to think about expanding our markets geographically because we were now part of Tetra Tech, but needed to keep our core operations profitable. Our private sector work was stable, as was our hazardous waste and transportation/highway work, but we needed more horsepower in our public wastewater work. This was also an area that Tetra Tech corporate was interested in expanding.

My solution was to hire Charlie Button, who I had known from my M&E days. He had been chief engineer of the Boston Water & Sewer Commission (BWSC) when we first met, and was chief engineer of the Massachusetts Water Resource Authority (MWRA). He was a good guy, and everyone knew him and respected him. He decided to leave the MWRA, and it was a real coup to hire him. I was thrilled.

Charlie helped us get some good work, but not the amount I was hoping for. It was the classic case of a terrific public employee not doing as well in the private sector. Despite my disappointment, I was very fond of Charlie, and we became quite close. We had an honorable parting of ways and shortly thereafter he became ill. I often had lunch with him in West Roxbury, where he lived, as he declined. He was a big guy, 6'3" and 240 lbs., but sadly wasted away as the weeks went by. He never complained.

Charlie was a legend in the local gin mills around town. There was a famous Dorchester Avenue pub crawl, usually done in the early spring around St Patrick's Day. There were dozens of bars on Dot Ave, and the plan was to start in the south, near the Milton line and work your way north toward downtown, having a beer in every joint. I think Charlie, who had a wooden leg, was the only person I knew who made it the whole way, although it might have taken him more than a day. I was lucky if I got halfway.

In addition to Charlie, I had hired Al Carrier, also from the BWSC, to help with the wastewater work. Poor Al thought his goose was cooked with me, right from the start. I went to his going away party from the BWSC and after dinner we went to a bar in Southie. I parked in the lot next to the bar, and when I left around 2 am, my car had been towed. Al thought I would hold it against him, but not to worry. Al was a terrific guy, and we became friends.

After Charlie left, Al took over the wastewater group, but we still needed some serious technical horsepower. I had the good fortune to hire Ding Fang, who had a PhD and was terribly underutilized in a small firm on the North Shore. He was fantastic technically and could explain complicated computer models to our clients. John Sullivan, chief engineer of the BWSC, said Ding Fang was the best he ever worked with. Al, Ding Fang, and I made a good team. We won several significant jobs with the BWSC in Dorchester and South Boston, and did their city-wide storm water management plan and federal permit.

I mentioned earlier that there were three people that stood out technically from the rest. Two I met at M&E, Abu Alam and Dave Langseth. The third was Ding Fang. He went to school, through college, in China and wanted to go back since many of his friends and classmates were running companies and wealthy. His wife, also a PhD in the biomedical field, wanted to stay in the U.S. to raise their kids. She won, and I was happy about the result. We had a design build job with Perini for the waste water treatment plant serving the Green Zone in Baghdad, Iraq. Ding Fang managed the design and kept Perini happy all the way.

I worked with a lot of good teams over the years. That's where the magic happens. At M&E there was the team lead by Jek for many years. At Rizzo I have mentioned the wetland team of Amy, Mark and Bernie, the private sector team of Brian and me, and the public sector team of Al and Ding Fang. While I was president I was always about 50% billable, mostly on private sector development projects. After Brian, my go-to private sector team became Steve Pavlovich and Natalie Brown. Steve was the perfect tech savvy civil engineer. You could talk through a project with him and arrive at a general sense of the final design. Steve would take that and come back with a complete set of plans. They were always better than I expected. Natalie would do all the calculations (water, sewer, drainage) and write the reports. We were a great team and they always exceeded my expectations.

In the end, the richness of your life is about the people you associate with and care about. When good people get together, and work together, the accomplishments far out weight the sum of the individual parts. To do this you need to trust people, bring a dose of humility to the table and be willing to give credit where credit is due.

5A. OUTSIDE FORCES

Before and during the time I was president of Tetra Tech Rizzo, there were two outside forces that shaped our destiny, and that of many others. The first was technology. Starting in the 1990s through 2010, the use of computers exploded in two ways. Communication was now done by email, 24/7, and the engineer/scientist could do most everything from his/her desktop computer, which soon became an iPad and smartphone. This eliminated the need for a lot of the traditional administrative staff (e.g. typing pool), and due to the 24/7 nature of things, efficiency improved. Instead of arriving at your desk and sorting through many messages that had piled up while you were gone, messages were returned in real time, all the time. You could talk in the car or type reports in your easy chair, at home by the fire. Engineers could also now produce a memo or technical report, with tables, figures, and photos, without any admin support.

Speed in communication has many blessings, but it can cause problems as well. There is a book called, *Thinking Fast and Slow, by Daniel Kahneman*. When you think fast it's based on instinct and although normally right, can be terribly wrong. Thinking slow brings in logic and more thoughtful analysis. So be careful with quick responses, especially if they are not necessary. Sometimes thinking things over in the shower produces better results that could avoid a costly lawsuit.

Remember, lawsuits cost money, but they can also have a devastating negative impact on the people involved. Avoid them by thinking slow and settle them as quickly as possible, even if you are right.

The second big technology change was in Computer-Aided Design and Drafting (CADD). With CADD, engineers could design a project faster, more accurately, and without the help of a draftsman. Just as important, if a change was made to a project with multiple drawings (sheets), all that was needed was to make one change and the program automatically made the appropriate change to all the affected sheets. Our chief draftsman, Pat Barret, who might have otherwise lost his job, learned CADD and singlehandedly dragged us into this new age. It was a game-changer and we were in the thick of the change.

One day early on in this process, we had a client in the office to look at the progress of one of his projects. I think it was the site plans for an office park. There were about 30 sheets in the design package. The client was expecting to see a hard copy of the drawings on a big layout table. Instead, Pat sat him down in front of an oversized computer screen to review the drawings. Every time the client wanted a change, Pat made it as he watched, and all the other drawings were changed automatically. After he was finished with Pat, the client and I sat in my office to discuss other project related issues. Less than a half hour later, Pat came in with a complete set of revised drawings that he had printed out and asked if the client wanted an electronic version sent separately. The client was speechless and essentially said he would never use another engineer, he was so impressed with the technology. Pat left with a big smile, and he deserved it, as he represented an extraordinary example of embracing change and not fighting it. Hats off to Pat.

These two technology changes boosted our efficiency significantly. We could now get more done in an hour than ever before, and the 40-hour work week was no more. I knew this instinctively, but the proof to me was our revenue growth vs. our headcount. In the 10 years from 1998 to 2008, our revenue doubled from about 20 to 40 million dollars/year, but our head count only went up by about 30%, 200 to 260 people. That's efficiency.

What happens when you can do twice as much work in the same amount of time, and you are billing by the hour? Who should benefit? The answer is everyone. But first you need more work for the same amount of people, to keep them busy. I'll get to how we got more work shortly, but first a few details.

The easiest way to improve your profit margin is to use a lump sum payment method, rather than time and expenses. For example, if your cost before efficiency was $100, and it's now $70 dollars, you can quote a lump sum fee of $90 and everyone wins or appears to win. If the client will not enter a lump sum contract, like many public agencies, you can add (legitimate) hours to your fee for QA/QC and principal (VP) time. This helps utilization for senior people, always a problem, as the situation with Liz illustrated. It also improves quality and should lead to fewer problems and potential litigation. You can also charge for computer time. Finally, if all the QA/QC and senior level hours aren't needed, they can be transferred to production time, especially if the project is going over budget. Like low utilization for senior people, losing profit on over-budget projects is often a problem. Efficiency due to technology gives you many ways to improve your financials and the quality of your work, providing you have enough work. So how did we get the work?

If you are more efficient your costs go down for the same about of work. This can make you more cost competitive in the marketplace and result in more work, potentially a lot more, especially if your competitors lag in being equally efficient. This factor can combine with the second outside force, a strong economy, to explode new business opportunities and successes.

The second major outside force that impacts every business is the health, or not, of the overall economy. As I mentioned, the first two years of the new century we recovered from the dot-com bust. After that, the economy boomed until the crash of 2008. In a booming economy it's much easier to look like a genius. During that time every part of our business was very strong. Private developers and public agencies were building like crazy, and the wars in Iraq and Afghanistan were fueling our work with the Corps of Engineers. All this provided the work we needed to feed our technically efficient staff, and we were making excellent profit margins.

In the last two years before the 2008 recession, we were the most profitable of the 30 operating units in Tetra Tech. This caused us to receive the ROI (return on investment) award at the annual meeting, the highest one given by Tetra Tech. Needless to say, with this award came a great deal of exposure and gave me a lot of goodwill points with Dan Batrack, our new CEO. Our bonus pool was over 1 million dollars each of those years, more in one year than all the years of Rizzo Associates. It's not fair to only look at cash bonuses, given that the ESOP turned into a bonus at the sale, but Rizzo Associates never came close to the profits we realized during those flush years before the 2008 crash. I had control over how the bonus pool was distributed, and certainly took my share, but because Rich and his international work were so instrumental in generating profits, I gave him a bonus bigger than mine both years. Brian also got bonuses close to mine because of his profitable private sector work. They both deserved them. It became crystal clear that joining Tetra Tech was a win for everyone.

We were riding high and wanted to take advantage of our influence, so Brian and I figured we could take our private sector development experience to other geographic locations. I'm not sure how all the pieces came together, but, on our recommendation, Tetra Tech became interested in buying Daylor Associates, with offices in Massachusetts and Florida. We were mainly interested in their operation in Southwest Florida. It was all private sector development work and things were booming. Sadly, the negotiations dragged on into the beginning of the recession and Tetra Tech killed the Florida part of the deal. It was a good decision, but we ended up not expanding our private sector work into other parts of the country. We did get Daylor's Massachusetts office, and its main asset, Bob Daylor. He was a good guy, knew everyone in the development business and help generate growth for many years.

By the mere fact that I was a president, I got a lot more attention from the outside world. I was on the board of ACEC Massachusetts and was appointed by then Governor Mitt Romney to the Goddard Council, which he established to promote STEM (science, technology, engineering, and math) careers starting in middle schools. I was very much in favor of this, and our company participated in hiring science teachers for the summer. They learned our business, by shadowing our engineers, and took their experiences back to the classroom. I thought it was great that by training one person (a teacher), you could reach many (students).

Unfortunately, I attended many Goddard Council meetings with mostly academics, where there was way too much talk about process, and the many available programs, with too little action. I also personally set up two scholarships, one an endowment at Lafayette College for a civil engineering student, and one at Cresskill High School, in my mother's name (the Althea Moore STEM scholarship) for a woman pursuing a STEM career at an accredited four-year college or university. Both are good investments.

I continued to be involved in WEF at the national level. My main contribution was looking at EPA's water quality regulations as the chairman of a task force on the subject. This was a high-profile project and I worked with engineers, scientists, and lawyers, mostly lawyers, to come up with a position paper that explained the process and recommended changes. We covered water quality standards and concepts like anti-degradation. It was cutting edge stuff, and I enjoyed the work.

The best part was going to the WEF annual conferences, which attracted 15,000 of the best people in the industry, in nice places like New Orleans. Tetra Tech always had 20-30 people attend, a booth, and a high class hospitality event for clients and friends. We worked, but also played. Charlie MacPherson was the Tetra Tech person who organized these events and seemed to know everyone who was anyone at Tetra Tech. She was a good person to have on your side. My first time in New Orleans as a Tetra Tech person, I went to a bar on Bourbon Street to meet other Tetra Tech folks, including Charlie. There was a band, and Charlie was dancing when I arrived. After a song or two, I asked her to dance. She was a terrific dancer, and we had quite a time. Thereafter, at other events, we often found a place to dance, from New Orleans to the Bellagio Hotel in Las Vegas. We had many a wonderful spin on the dance floor and became friends. She was a quality person, as were many others at Tetra Tech.

5B. INTERNATIONAL ADVENTURE.

We were riding a wave, better than taking drugs, but I was actually getting bored. I had done everything I could do and was looking for something new. Doing the same old stuff year after year didn't get me excited anymore, although the money was good. Even working on projects was getting old, so I went in a direction that seemed natural, international work. Tetra Tech's international work had expanded somewhat after Li-San retired, but it was still mostly with USAID or the military. This was considered safe, getting paid by the U.S. government, and Tetra Tech was a first-tier beltway bandit. They had 1,000 people in half a dozen offices in the Washington, DC area. This was fine, but I wanted to go after non-U.S. funded work.

I helped Dan with an acquisition of a large Canadian firm that had an office in India and with another large US firm, ARD, in Burlington VT, that did USAID work exclusively. I got to know these two firms after I got involved in the India operation. ARD did over 100 million dollars in revenue, with only 50 or so people in Burlington. Most employees were local people in the host countries. For example, there were 300 people in Bogota, Columbia, and all but a few were locals. Their main project was to help the government get farmers to grow something other than poppy plants, the precursor to the heroin found on US city streets.

When the sale to Tetra Tech was announced, the ARD staff in Burlington went crazy thinking that they were being taken over by the military industrial complex. Most were liberal tree huggers, good at what they did, and used to their independence. I was asked to go to Burlington with the senior folks from Pasadena, to calm the natives, by telling them how wonderful out transition was. I could be honest and told them not to worry. I used the 80/20 rule, or in our case it was 90/10. 90% of what they had been doing would not change, and only 10% would. Of that 10%, most was good change. It was true, good PR, and it apparently worked. When I next went to Burlington, their new president, Jan Ammon, took me to dinner at a fancy restaurant and thanked me over some expensive wine.

With my newfound clout as president of the most profitable Tetra Tech unit, I could take some chances and get away with it. My first effort was in the Middle East and the second one was in India, as you will see in more detail later. But first the Middle East. After all, Bill had spent 1 million dollars going after work in Poland, as previously noted. Why shouldn't I try somewhere else?

I knew next to nothing about non-U.S. funded international work, but Tetra Tech was a $2 billion, publicly traded company that gave me credibility. Tetra Tech bought a mid-size company in Florida, Ardaman Associates, which specialized in working for the phosphorus mining industry. They were mostly geotechnical engineers, and they worked all over the world, following the industry. The president was a great guy, Anwar Wissa, born in Egypt. He had been a professor at MIT, married a Massachusetts girl, and was lured away from MIT to run the company when the founder, his old professor, retired. His wife stayed in Cambridge, while Anwar commuted back and forth to Florida. Anwar had come to the U.S. for a PhD at MIT. Ardaman had a small branch office in Cairo, set up by Anwar. When Li-San bought the company, he excluded the Cairo office, so Anwar became the sole owner of this little outpost. He naturally had a professional and personal interest in growing the office.

I would see Anwar at quarterly meetings for the unit leaders, and we got to talking about our mutual interest in overseas work. His unit was not doing as well financially as we were, so it was agreed that I would take the lead (and risk) in testing the waters. We also agreed that Qatar was a good target country. Qatar wasn't Dubai, where there was too much competition, but it had lots of money and wanted to be the next Dubai. It was also in the running to host the 2022 soccer World Cup.

As I was doing some due diligence, it came to my attention that Patton Boggs was the only U.S. law firm that had an office in Doha, the capital. It also so happened that my dear friend and fraternity brother, Jim Schwarz, was working for Patton Boggs. The stars were aligning and Jim set up a meeting in DC for me to meet their key players in the Middle East. They agreed that Qatar was ripe for foreign intervention, not surprising given their office there. To make things interesting, the head of their Doha office was a woman, in a Muslim country! We put Patton Boggs on a retainer, and they agreed to meet me in Doha and set up some introductory meetings.

I planned a two week trip to the Middle East, one week in Doha, five days in Cairo, and a one-day stop in Beirut. Anwar's Cairo office manager, an engineer, was to meet me in Doha after a few days and we would be a team for the rest of the trip. I did this mostly on my own, since Tetra Tech, bless their heart, didn't ask many questions and I didn't give them reason to. After all, the Cairo connection wasn't even part of Tetra Tech and Anwar was footing that bill. Best of all, I had a local contact I could trust and who could interpret the local magic. So off I went, mostly optimistic because I was naive and had no real idea what I was getting into, but I had Patton Boggs and my man in Cairo for backup.

When I got to Doha, I met with the folks at Patton Boggs and they had indeed set up a meeting with a local guy, who had some distant family connection with the royal family. He was a cousin to some bigger fish, as I remember, and only about 40 years old. For some reason we hit it off right away and bonded, not that I had too many other choices. He took me under his wing and started to set up meetings with folks closer to the King and Queen. We had 5 or 6 meetings sitting on the floor in ornate rooms drinking lots of tea, talking with various levels of the royal family. This was terrific and even though we had no formal agreement with my new buddy, everywhere we went I said he represented us. This was the right thing to say, as he was thrilled to be associated with a large U.S. firm, traded on the Nasdaq stock exchange.

In Qatar the only way to work was to team with a local firm, who technically owned 51% of the business. So the purpose of our meetings was to flush out the best teaming partner, with my friend as the go-between, our agent. Keep in mind that Qatar was booming with all sorts of real estate and infrastructure development, much anticipating the 2022 World Cup. My interest was in the infrastructure work and the soccer stadiums. After all we were the engineer of record for the Patriots stadium.

All the local folks wore typical Arab clothes, so I stood out like a sore thumb. One morning I met our guy for breakfast, and he was late. He arrived and looked a bit stressed, so I asked why, now that we were buddies. Turns out many Arabs have more than one wife. He had two, which was the good news, but the night before they both had demanded his attention, and he was exhausted. He told the story rather sheepishly, and we both had a good laugh.

By the end of the week, we were about ready to take Qatar by storm and I was quite happy with how far I had gotten. The only problem was I couldn't make any final decisions, only report back to Tetra Tech for their approval. As a result of our talks, we had narrowed down potential partners but made no commitment. It also became clear that the best way to establish yourself in the Kingdom, was to make a sizable contribution to a cause near and dear to the royal family.

The Queen, who was progressive and very accomplished, was the force behind the building of a major medical center and medical school. Some would say she was the power behind the throne and if you had her backing, things would go smoothly. She was in partnership with the Rockefeller Foundation and Carnegie Mellon for the medical center and school. The facilities were soon to open, and to endow a chair for one of the medical professors was $100,000. Making this contribution to endow a chair was above board, done publicly, and perfectly legal.

I thought this was a terrific way to impress the queen and instantly become a player. At 10% profit, which I thought was low for working in Qatar, you only needed 1 million dollars of revenue to get your money back and I thought you could do $10 million/year for the next 10 years to the 2022 World Cup. What a deal, but first I needed to complete my trip and get back to make my case to Tetra Tech. I was excited and wanted to get back, but first Egypt.

I stayed in a wonderful hotel on an island in the Nile that was a converted palace of some long-forgotten sheik, expanded with a modern wing. I didn't do much work and was mostly a tourist, chaperoned by Anwar's local guy. Every night the men sat in the open-air courtyard drinking and smoking water pipes until dinner was served around 10 pm. Needless to say, the next day didn't start until mid-morning. One night one of Anwar's sons, who had relocated to Egypt, stopped by for dinner and we had a fine time. A good guy, who wanted me to travel to the Red Sea and go scuba diving with him. Some of the best diving spots in the world, but not that trip.

Part of the reason I wasn't going anywhere was I got a severe case of gout in one of my ankles, which made walking difficult. I didn't know it was gout at the time, and thought I had a sprained ankle. In any case, it hurt like hell and gave me a good excuse to head home. I wanted to visit Beirut, but it wasn't in the cards. The only productive thing I did in Cairo was to write a short business plan for Qatar, which I planned on presenting to Dan upon my return. I sent a copy to Anwar for his input, and he quickly got back with his approval.

Once I got home and my doctor diagnosed my gout in 5 seconds, I got medication that worked in a flash. I was now in good spirits to meet Dan and make my Qatar case with Anwar by my side. The next quarterly meeting happened to be in Palm Springs, CA, two weeks out. Good timing and a nice venue. After breakfast one day, sitting on the hotel patio in beautiful Palm Springs, Anwar and I gave a short presentation to Dan and his CFO, Joh King. I can be pretty persuasive and made a good case that if we wanted to play in international waters, we needed to jump in and not just pussyfoot around the edge. This is where I proposed the $100K donation to the Queen's medical project, which I thought was a good investment given all the work expected in the next 10 years. The $100K caught Dan a bit off guard and he laughed. Joh choked and spilled his Bloody Mary.

After Dan stopped laughing, he was nice enough to say we did a good job of developing the proposal, but he decided not pull the trigger. I believe the reason he killed the deal was because he was not comfortable with non-US international business, as this was way outside of Tetra Tech's comfort zone. Remember, Dan was a protege of Li-San and his business model, and he was new to the role of CEO. Dan also made the point, a good one, that the $100K donation would not be the last request made of us, before we started to make some money. He was probably right, but I think it would have been a success in the long run. We will never know, but Qatar did get selected for the 2022 World Cup, and they spent billions preparing for the games!

Anwar and I were disappointed, and I had to call my Qatar buddy and tell him the bad news. We kept in touch for a year or two, but nothing happened. It was a good experience for me, and at some level I was relieved I didn't have to deliver on my optimistic proposal. I focused back on continuing to keep Tetra Tech- Rizzo as profitable as we had been, but I couldn't stay focused long, as India came into the picture.

It is important to point out the benefits of working for a big company. But these benefits don't come automatically, you need to cultivate them. I'm thinking primarily of cross selling services and teaming with other Tetra Tech companies to get bigger jobs. To do this I made a point of getting to know the other unit leaders, so they would think of us when going after new work. I also offered to pay our share, and more, sometimes much more, of the new business costs. This got people's attention, and they remembered it. It was like planting seeds that you hoped would grow, and they did. It also worked the other way, when we had a local job that required expertise we didn't have. You teamed with another unit that did and won the job. This happened several times with water quality modeling projects where we used the DC folks who had the best models in the world and knew how to use them. The client agreed and selected the Tetra Tech team. You have to give something to get something, and unfortunately most people aren't willing to make that leap.

When I left for India, Brian became the third president of the company. I like to think of it as a sandwich. Bill is the bottom piece of bread, Brian is the top piece and I'm the good stuff in the middle. There was never any real question about who would be the next president. I had hired Brian from M&E and we worked together for our entire time at Rizzo/Tetra Tech. He had earned it and there really wasn't any competition. He was great with clients and despite his Irish personality, he was at his core a conservative engineer. It's hard to put a value on something that never happened, but Brian's work never got us into a lawsuit. Thank you Brian.

When Brian took over, he asked me for my advice on how to run the company successfully. I was flattered and gave him three suggestions, all related to people. The first two were to keep Rich Sperandio and Ding Fang happy, as they were irreplaceable. The third was to build bridges with the presidents of the other Tetra Tech companies, as they could be your best future clients or teaming partners. The message was to use all of Tetra Tech to your advantage, not just your local capabilities. I don't believe Brian hung this advice on his office wall as a reminder, but whatever course he took, he has been successful.

It was a good transition, and my only regret was the company was split into two shortly after I went to India. Rich Sperandio's federal operation became a separate operating unit, under his full control, something he well deserved. Brian kept the rest of the company, which was essentially Massachusetts focused.

CHAPTER 6
THE FUTURE

Before we get into the India affair, I will take stock of the current situation (2025) and how it might affect the engineering consulting business going forward. I hope you understand, from what you have read so far, that the heart of consulting is the people, and that is the good news. The bad news is that the national environment, in which the business operates, is being poisoned.

As I have mentioned I had a wonderful life/career, working with great people. Maybe it is something in the DNA of civil engineers working in the environmental space, but I have been blessed. I am on the Board of a civil engineering company, my college fraternity, and still do pro-bono work, including in the environmental justice area. The young people I meet in college and professionally are hardworking, have integrity and maintain the professional standards I was taught. I'm optimistic about the future of the Civil/Environmental profession.

There are, however, policies being implemented in Washington DC that will cause serious long term problems, not only for civil engineers, but for all of us. These are being created by the current federal administration. The first has to do with the discrediting of the Environmental Justice movement. I believe the evidence is in and it is clear that poor communities, black and white, but mostly black and brown, have seen a disproportional share of adverse environmental impacts. Highways and industrial developments, for example, have displaced and isolated poor communities and inflicted air, water and land pollution in far greater proportion than in more affluent, mostly white communities. Not to recognize this and try to redress some of these inequities is a moral failure.

The typical argument against environmental justice is the one used in so many other similar cases. "It is a free country and any one can achieve success through hard work". The implication is that the reason black and brown people are in poor neighborhoods, is that they are ignorant and lazy. I agree with the first idea that this country offers great opportunities for those who strive to take advantage, but not everyone is given a fair playing field to success. Black and brown folks don't get a fair break in financial markets, housing, medical care and education. Until these issues are looked at with clear eyes, something needs to be done now, to mitigate decades of environmental impacts. Controlling air and water pollution is not so complex, The same is true about shutting down a rock crusher at night so nearby neighbors can get a good night's sleep. It mostly takes political will and money. The US has little of the first and a lot of the second, but poorly distributed.

There are many statistics that illustrate income inequality, and the fact that it is getting worse, but there was one statistic that stopped me in my tracts and sums up the problem. In the city of Boston the difference in life expectancy between people in the Roxbury neighborhood (poor and 69 years) and the Back Bay neighborhood (wealthy and 92 years) is 23 years as reported by Boston's Public Health Commission, as of 2023. That's right, 23 years. If that doesn't require asking some hard questions, I'm not sure what does.

Our fellow citizens who live in these environmentally compromised neighborhoods, are the workers who propel our national economy. If they cannot work because they need medical support, or cannot afford child care, we all lose. It would be cheaper in the long run to keep them healthy. Put aside the moral argument, and it's good business practice to clean things up.

The Trump administration has created a short-cut for industry to avoid clean air rules, by simply sending an e-mail to EPA requesting an exemption. No documentation is needed, and the approval is based on the 'interest of national security', a familiar excuse use by the administration to avoid due process. Exemption have been granted to coal fired power plants, steel mills, oil refineries and my favorite, a copper smelter in Arizona discharging lead, arsenic and chromium, owned by Freeport-McRoRan.

There are only two copper smelter plants in the US. The other plant has installed air pollution equipment to meet standards. The plant in

Miami Ariz. has not and continues to pollute the community, many of whom work at the plant. Over the years I have often questioned certain EPA regulations and red-tape, but this casual exemption process will create short and long term health issues for the people in Miami, especially the children. This given that Freeport's annual gross profits last year was $7.7 billion, clearly enough to cover the cost of installing and operating the air pollution equipment, estimated between $60 and $300 million, depending on who is making the estimate, the regulators or the company. It's the age old story of maximizing profits, versus taking responsibility for the pollution you cause. The administration clearly cares little about the need to run responsible businesses.

The civil engineering profession is seeing larger and more complex projects: floating cities, space vehicles, high speed rail transport, fusion energy and the list goes on. These mega projects will require sophisticated project management and technology, including artificial intelligence (AI). The people needed to build these projects will need to be top notch, as well as those people who develop the tools to make them possible.

You will recall that two of the three smartest people I have worked with are from Bangladesh and China. My Chinese friend may finally decide to go back home, because he will be afraid to visit and not be able to return to the US. We desperately need this talent and the current administration is driving them away. We are already starting to see a brain drain, damaging one of American's greatest competitive advantages over the past 75 years. It will severely diminish America's standing in the world, because intellectual capital was what made America great. It gets worse because basic research is being defunded to pay for a tax break for the wealthy. Other countries are luring our best and brightest away with research opportunity we should be supporting. This research is our future, and the current government policy to defund it is just plain stupid.

It is not only the smart folks who we are losing. In order to build large projects you need labor. Talk to any general contractor, or for that matter, your neighbor who is involved in a home improvement project. They will tell you that undocumented immigrant labor is necessary and the immigrants are some of the best workers. They are not criminals, they love their families and they pay taxes. They are us.

We are impeding immigration, and the immigrants who are here are being rounded up and deported to countries you have never heard of, or choosing to leave of their own accord. All the while there are vacancies for 400,000 good paying manufacturing jobs, and this may rise to over 2 million in 5 years, The Trump folks are apparently worried that the country will have a minority white population in a few years. This is what I call a version of 'white supremacy', which is an ugly philosophy. When will we accept that all of us have the same ancestry and DNA. The world's population is projected to peak in the next 30-50 years, and the subsequent decline may be dramatic. At that point we will be begging people to come, regardless of their color.

The other thing that is devastating to our work force is our lack of investment in seeing that all children are properly fed, housed, educated and have access to medical care. This is the best investment the government can make in our future, and it can be done. Actually, it was done during the pandemic, in terms of food availability. We have the resources, not the will. This is not socialism. This is good economic policy, with a large return on investment.

Climate change is another area in which I strongly disagree with the administration. It is hard to understand how a policy that supports fossil fuels at the expense of renewables is justified, except for the political clout and money from the oil and gas companies. These companies have known about the problem for decades and have continued to pollute at no cost, leading to their enormous wealth and power. Even Milton Freedman agrees that companies should be held accountable for the cost of the damage caused by their pollution. The tax bill recently passed by the republicans rewards continued zero cost pollution by the oil and gas industry. Maybe ExxonMobil should pay the billions of dollars it will cost to construct a dike around lower Manhattan Island to protect against sea level rise. After all, they made about $35 billion in profits each of the last two years. They could pay for the dike with one month's profit!

Most people, except for Trump, believe the planet is warming. Some argue that the cause is unknown. The scientific community is pretty clear that burning fossil fuels are the main cause. Most other theories (e.g. sun radiation, natural variations in the carbon cycle) have been studied and are either accounted for or ruled out in making projections. I have done a lot of computer modeling and know that any prediction has a level of uncertainty. That doesn't mean the results can't be used as a guide in making energy policy. Even if you are not convinced by the weight of evidence that fossil fuels are the problem, it is still good economic policy to support renewables.

The wind and solar markets are some of the strongest current drivers of our economy, many times bigger than oil and gas. Companies I have been involved with have done significant civil engineering for the wind and solar industries. The energy produced is cheaper then oil and gas in many states, so make no mistake the cost of energy will go up as a result of this tax bill and the energy will be dirty. Just as bad, we are giving China a free pass to dominate these markets. Also consider the possibility of being wrong and we ignore the problem. The future is then very bleak for those we pass it on to. It is already heading in the wrong direction at unusual speed.

It is also easy to think that some magical technology will save the day. Think again, and do not bet the future on a hope. However, there is one existing technology that is my personal favorite to address the climate issue: nuclear energy. It is viewed favorably in the tax bill, a rare positive, and could solve the problem in a safe way. I may be biased in that civil engineers have a large role in building nuclear power plants. Unfortunately, public opinion has been poisoned toward this technology. The oil and gas industry know this, so it felt safe allowing it in the bill. I like to point to a bumper sticker I saw years ago when Ted Kennedy was running for his senate seat: more people died at Chappaquiddick then at Three Mile Island.

In short, the climate/energy policies of the current administration are madness and will be viewed as such in the coming years, but hopefully not before it is too late.

There is an argument that the current administration makes about the federal government being too big, inefficient, regulation-happy and a drag on the free enterprise system. I'm sympathetic to this, as any large organization can occasionally use a kick in the pants, but destroying institutions with no plan to rebuild is crazy. Institutions can be terribly frustrating, but a complex functioning society needs institutions (e.g. CDC, IRS, Federal Reserve, media, courts) that can be trusted. There are many laws that govern our collective behavior, but the glue that makes a democracy work is trust, in our neighbors, institutions and the government, writ large. The current administration, by claiming they are the only truthful voice, is flirting with Fascism by destroying trust.

You can trust someone, or an institution, even if you don't agree with them, providing there is a willingness to engage in respectful conversation on different issues. What does Trump do when he doesn't agree with you? He calls you names in a disgraceful, school yard manner. The names are too demeaning to print and illustrated by one of many articles titled: "598 People, Places and Things Donald Trump Has Insulted". You can't build a humane society like this.

In our system, economically, there needs to be a balance between uncontrolled capitalism on the one hand and socialism on the other. Just as politically, there needs to be a balance between the three branches of government. That is when we are at our best, but the current administration is destroying this balance both economically and politically. This will end badly for all of us. A good start in turning this chaos around would be in telling the truth.

The second half of this book is about my international work in India. Although I talk about how Tetra Tech wasn't interested in international work, I believe the big US civil engineering companies will need to work internationally, to prosper in the future. This is because they need the new business and they need to learn from professionals around the world. Right now, the current administration is building a wall (literally and philosophically) around the US. There is not a country that trusts us and for good reason. Our policies, if you can call them that, are ever changing. If Trump happens to get up on the wrong side of the bed, he can call your country a loser, or a lot worse.

Trump has the attention span of a child. More sophisticated folks think ahead, analyze difficult problems and have strategies. Would you want to work with someone who makes a deal and changes it the next day, for no apparent reason. As a businessman, and one of the worst in the last 50 years, no one would work with Trump, because they could not trust him and/or get paid. Same mode of operation as president. Think of how hard it is to piss off the Canadians, and they are pissed off big time.

Going forward we need partners, just as you need a good team to get an engineering project done successfully. We have no partners and we have antagonized most of the world, except for Netanyahu, who is destroying Israel's standing in the world, just like Trump. We have cut off foreign aid (USAID), the most cost effective way to project our influence around the world. Even the senior military folks will tell you that. This aid has saved millions of lives, and now it will not.

Estimates indicate that USAID saved 93 million lives over the past 20 years. Current information indicates that withdrawing this aid has already caused 600,000 deaths, 2/3 children. By 2030 it is estimated that 14 million deaths will occur, 4.4 million children, and that doesn't include support for education and other social services. These numbers are truly breathtaking and profoundly sad. In the future these wrong-headed policies will prevent US companies from getting work and being trusted. This is not making America great, this is losing our moral authority and crushing opportunities, for years to come. Mothers and children, here and the world over, are dying because of these policies. Think about that the next time you vote.

Sadly, China's and Russia's main strategic goal is to weaken America's standing in the world. They couldn't be happier that Trump is president and is unwittingly working on their behalf. The policies of the current administration in DC (immigration, energy, soft power, R&D, children's welfare, higher education, press relations, rule of law, tariffs, respect, and the list goes on) are a betrayal of the American people and will be a calamity for future generations.

A foreign policy that will sell out Ukraine, ignores our allies, and breaks up the world into spheres of influence that history teaches us is a recipe for future wars. It denies our role in keeping peace since WW II, a key ingredient, and excellent investment, that allowed for our current success and prosperity.

Trump was impeached twice in his first term and I think there is a strong case for a third impeachment, and the charge would be murder. If this sounds far-fetched, consider the following indictment, which doesn't include the loss of life on January 6, 2021 and the sedition that preceded one of the darkest days in American history.

1. Killing of alleged drug smugglers at sea, and lying about the reason, which was not drugs, but Venezuela's oil.
2. Elimination of USAID and the funding of food and medical support to nations around the world, that have saved millions of lives.
3. Advice from the Department of Health and Human Services that eliminates the requirement for some vaccines and confuses the public about the need for vaccines in general, based on no scientific evidence, and against the advice of every major medical association.
4. Reduction in funding for medical research at major universities and hospitals, focused on the next generation of health cures.
5. Reducing funding for food and medical insurance to support lower income families, mostly children, in favor of tax breaks for the wealthy.
6. Exemptions, granted by EPA, from air pollution rules for coal, oil, gas, steel and copper smelting industrial facilities.
7. Calling climate change a hoax, and declaring war on renewable energy sources in favor of fossil fuels.
8. Abducting persons from neighborhoods by masked ICE storm troopers, without due process, and transporting them through internment camps to third world countries where many will perish.
9. Pardons, based on corruption and personal grievances, of people who will continue unlawful practices and promote political violence (Proud Boys and Oath Keepers).
10. Failure to address gun violence with reasonable gun laws that are needed to address a serious public health issue.
11. A foreign policy that will sell out Ukraine, ignores our allies, and breaks up the world into spheres of influence that history teaches us is a recipe for future wars. It denies our role in keeping peace since WW II, a key ingredient, and excellent investment, that allowed for our current success and prosperity.

I do not like to end on a sour note, because I do have hope in the younger generations. They will bend the arc of history toward justice, as MLK said, but they will need support. The countries that realize this will prosper. India has the advantage of having the largest youth population in the world, which is the raw material of success, and why I am optimistic about their future.

CHAPTER 7

INDIA INTRODUCTION, 2008-2012

As I mentioned, I started to get interested in India around 2008. At that time India was part of another operating unit and did work almost exclusively for USAID. This was the standard Tetra Tech model for international work. The local guy who ran the office, Satish Malak, was high energy and had started to win local work. Unfortunately, one of the jobs had run into payment issues due to a change in the local state government, our client. Partly because of this, Tetra Tech was planning to shut down the office. At that point, I stepped in and offered to take over the operation. Normally, the answer would have been no, but because I was looked upon favorably, due to our excellent financial performance, I was given the green light to take over the operation and see what I could do. Dan Batrack warned me that, in his opinion, I had almost no chance of success. This, of course, sealed my determination to take on the challenge.

Looking at it this way, I was expected to fail, and not cost the company too much money. With even modest success, I could be a winner. Keep in mind, this was during the 2008 financial meltdown and looking for work outside the U.S. seemed to me a brilliant idea. I was in Tokyo at lunch negotiating with Tokyo Engineering about teaming together and going after a large job in India, when we heard about the Lehman Brothers failure, the Wall Street chaos, and the beginning of the Great Recession. Also, I was at the end of my full-time career (I'm still working part time) and I really wanted something different. Little did I know.

By the time I got interested in India, Dan had taken over as CEO of Tetra Tech and added a layer of management between himself and the 30, now almost 40, operating units. No more calls from Li-San. He divided the units into three groups and appointed three people to lead these units, reporting directly to him. One new hire, Doug Smith, became my boss. He was a great guy, with engineering and law degrees. He also came from another large consulting company, Black & Veatch, where he had headed their international operations. This seemed perfect for me, as I was entering the international arena with little experience. He immediately hired Rick Wilson, who had previously worked for him doing international marketing. I was thrilled with the support. Rick had done a lot of work funded by the Japanese government and had contacts there.

It just so happened that Japan was focusing much of their international development work in India, funding infrastructure work in transportation (roads, subways, and airports), energy (power plants), and water/wastewater (drinking water and pollution control). The idea, as best I could determine, was to improve India so it could compete with China. Although much of the money to India came from Japan, this was coordinated with the U.S. at the highest levels, as geopolitical chess.

We were planning to use Rick Wilson's contacts in Japan to team with Japanese companies and pursue work in India funded by Japan. The money came as loans and in order to be successful, you needed to be teamed with a Japanese firm, even though the Indians managed the selection process. In our case, our teaming partner was Tokyo Engineering. They turned out to be good to work with and honorable business partners.

At the beginning, Tetra Tech had little or no experience with large international infrastructure projects, so we had to be a subcontractor to the prime consultant, in this case Tokyo Engineering. The first job we won in 2009 was a large wastewater collection and treatment system for the city of Amritsar in Punjab Pardesh (state). Amritsar is the home of the Sikh religion and the Golden Temple. Nice place in northwest India.

Amritsar is only about 15 miles from the Pakistan border. The city of Lahore is another 15 miles past the border. When the British left in 1947, they drew a hasty border that carved out Pakistan (east and west) as Muslim countries and left the remainder of India as Hindu. At independence, there was mass migration and killings, as millions of Muslims moved to Pakistan and Hindus to India. Satish and I met with his friend, whose family (Hindu) migrated from Lahore to Amritsar in the 1947 chaos, before he was born. They had established a company in Amritsar that produced activated carbon, and was very successful, but the generational scares were still there. The book 'Freedom at Midnight' is an excellent telling of how India gained it's independence. I recommend it.

Once we had the Amritsar project experience, we could argue for a larger role in the next job, which was a huge wastewater/drainage project in two cities in the state of Odisha, on the east coast of India. Each city (Bhubaneswar and Cuttack) had a population of about 500,000 and growing. We won this job as a consortium of firms: Tokyo Engineering (Japan), Tetra Tech (U.S.) and Egis (French).

As I mentioned, Tokyo Engineering was a great partner. Not so the French. They were arrogant and wouldn't agree to the simplest thing without a fight. I spent so much time arguing with them, I wanted to scream-- and did. My greatest explosion occurred when they sent a senior political scientist to reorganize the department of public works. First, he didn't accept that he was reporting to me and not the boys back in Paris. Not a good way to start. Second, we had organized a meeting to introduce him to the chief minister. I told him it was a happy meeting just to get the minister's ideas and to ask questions. After all, they had spent the last two years studying organizational alternatives and had essentially settled on an approach.

This meeting should have been simple, but he announced to me that he didn't like their approach and was about to tell them so. One of our local employees, a good guy, had been the chief engineer for public works and knew where all the skeletons were buried. I sat the three of us down and explained in no uncertain terms that our Frenchman was not, I repeat, not, going to tell them they were all wrong, after two years of work. We were going to save that bombshell for later, if at all, and the former chief engineer backed me up. So the three of us went to my "happy" meeting and the Frenchman proceeded to tell the minister their plan was a mistake. I couldn't believe what I was hearing and cut him short. The whole city could hear me chewing the Frenchman a new asshole on our ride back to the office.

I sent a blistering e-mail to the French guy in Paris who was their principal in charge, explaining how enraged I was at their performance in front of the minister. I also had talked to one of our management guys who didn't even understand what the Frenchman was talking about and thought the locals were on the right track. I also sent this to Paris. When the jerk came back for his second tour, he was very nice to me and got on board with the program. The moral of the story was not to team with the French, ever again.

On the Odisha job, Tetra Tech was first among equals, since we were to supply the project manager. As mentioned, this was something the French had a hard time getting used to. The scope of work included engineering, environmental documentation, and setting up a structure (Department of Public Works) to manage the new infrastructure. We were also to stay involved in construction management. The initial fee was the equivalent of 10 million U.S. dollars over two years, with more to come when construction started.

When the guy I picked to be the PM backed out at the last minute, I saw an opportunity. I wanted to relocate to India, get in the trenches, and help develop the business, but my U.S. salary was too high for the local operation to absorb as an overhead cost. If I became the PM for the Odisha cities, I would have billable work to offset my overhead cost. I figured I would work about half time on the project and half time on other things. Also, my living costs would be paid by the project and I could live in Bhubaneswar, which was so much more civilized than Delhi, with its terrible air pollution and traffic.

I had to spend time in Delhi, Amritsar and other places quite often. This was made easy by the excellent domestic aviation industry. The local airlines were reasonably priced, reliable and all major cities had new airports or they were being built. My main complaint when traveling was that locals, mostly older women, had no concept of personal space. You would be standing in line at an airport and a woman would cut in, drop her bag on your foot, and spill your coffee. Not even a 'I'm sorry' would follow.

I signed up for the PM job in Odisha and lived in India for about 6-8 months per year, for the next 2-3 years. Before we dive into those years, a few words about India's culture, business, and politics. Some words I didn't/couldn't say while in India and employed by a U.S. publicly traded company.

7A. CULTURE

The Indian people are the most hospitable in the world. Sometimes embarrassingly so. They are keen to anticipate and address your every need and make you feel part of their life and family. Satish and his wife certainly made me part of their family in every way, and their kids called me Uncle. Satish confided in me on personal as well as business things. One time he mentioned that his wife was acting strangely, and he was worried. I suggested that it could be menopause and that she should talk to her doctor. A few weeks later he thanked me for the suggestion.

In India, family and education are everything. With all the millions of young bright kids getting educated, India will be a world power long into the future. They do need to come to grips with the position of women, by breaking down cultural and religious barriers that hold women back. Women are the bedrock and hidden strength of India and this needs to be realized.

7B. BUSINESS

There is a bit of Jekyll and Hyde when comparing/contrasting culture and business in India. The culture is generally open, warm, and inclusive, while business is secretive, manipulative, and cutthroat. In business everyone is always plotting to get privileged information, the latest gossip, and the upper hand in any kind of negotiation. In this sense it's hard to trust anyone and makes for a stressful business environment.

In addition to the backroom shenanigans and drama, there was open warfare. This was often done on the pages of local newspapers. India was in the Yellow Journalism phase, as the U.S. was 100-150 years ago. The many newspapers had a thirst for readership and were willing to print anything. What better than a local scandal? Outrageous stuff was printed, often without verification. The same trend we see today in some of our so-called news outlets.

When we won the job in Odisha, our main competitor was pissed off. I knew the lead guy because he was on our team in Amritsar and had done some shady things there. He was from Nepal and had an engineering PhD, but no moral compass. He sent an anonymous article to the local newspaper accusing us of bribing city officials, and that my resume was a lie. Needless to say, this caused some trouble which took weeks to sort out. Our client knew the accusations were bullshit but needed to go through many steps to clear the air, including a notarized letter from me, co-signed by our CEO, verifying that my resume was accurate.

Speaking of corruption, a claim of corruption in India was almost always true. In the case noted above, we focused on refuting the claim that my resume was a lie and ignored the broader claim of a payoff. The plain fact: nothing got done in India without some type of payoff. Doug Smith and Rick Wilson had been playing this game for years and there was a simple formula. You hired a third party agent, in our case Mr. Watanabe, with his Japanese connections. He had worked for the Japanese equivalent of USAID and knew all the people. He was retired, in his late 70s, and had been a kid during the U.S. bombing of Tokyo late in WWII. We had many conversations and I liked him at some level, but he was a complete chauvinist. He looked down on anyone not Japanese, especially the Indian folks. Americans were just OK. When he traveled, he carried well wrapped rice cakes and ate them instead of the local food, even at the best restaurants. Interesting guy.

We had an agreement with Watanabe to act as our agent pursuing projects in India funded by the Japanese government. He was paid with a monthly retainer and got a fee, 2% of any contract we won. The agreement stated that he needed to follow U.S. laws, including the Foreign Corrupt Practices Act. This contract was to protect us against any funny business Watanabi pulled in private. He never told us, but my guess is he took some of his retainer and passed it along to key decision makers in both Japan and India, influencing their decisions. The only time he hesitated was in Goa, on the southwest coast of India, when he realized the chief minister was also a drug dealer and a relationship with him might not end well. I couldn't put any of this stuff in my blog, because e-mails often become the smoking gun in a corruption trial and I didn't have proof of what was actually going on. Of course, plausible deniability was the whole point of this setup.

When you won a job, you also needed a local agent. Watanabi set this up during the selection process and then he was mostly out of the picture. We had a great one in Odisha and a crook in Amritsar. It all added a great deal of stress, given that the Foreign Corrupt Practices Act was real and there were prosecutions under it with fines and jail time. Since I was a director of the India company, I was exposed. Now that Trump has said the Act will not be enforced, it's going to be the wild west, even more so.

I mentioned earlier that we had a receivable of almost $1 million by the Uttar Pradesh, from before I got involved. For me the holy grail was to collect this money and make our financials look terrific. I kept reminding Satish about this and after a new chief minister was elected, things looked promising. He thought we had a deal, but the locals wanted more money upfront. I said that was not happening, throwing good money after bad. Soon after, Satish took me aside and explained that he had taken $10,000 of his own money (his family was wealthy) and given it to our mailroom boy in a briefcase to take to the local decision maker in Uttar Pradesh. This was to seal the deal, but unfortunately, while walking through security at the airport he was stopped and the $10K was confiscated by the police. Satish told the police it was his money (correct) for a real estate deal (not correct). He wasn't too upset about losing the money, but he was upset that the deal collapsed. We never got the million dollars and I didn't tell Tetra Tech about the incident at the airport. It wasn't their money, but it was being carried by a Tetra Tech employee, and I should have come clean.

This stunt by Satish and the accusation that I doctored my resume, among other similar situations, created a number of very stressful days for me in India. Satish never told me of his many behind the curtain deals, following the concept of plausible deniability, and protecting me. Unfortunately, at any time, a case could have been made accusing us of violating the U.S. Foreign Corrupt Practices Act. I remember one case made against another major U.S. company, who paid a large fine and considered it the price of doing business. They were found guilty because of e-mails that were discovered, explaining exactly what happened. The only way to justify the corruption was to say that everyone did it, which was true, but little comfort. Because everyone did it, no one wanted to blow the whistle and upset the gravy train.

Once I understood how business was conducted and that it wasn't going to change, I told Doug Smith I thought we should close down our local business and only work for internationally funded projects (US-AID, Japan, Asian Development Bank, World Bank). It wasn't only the corruption, but also the problem of getting paid, as illustrated by the situation in Uttar Pradesh noted above. In real time, day to day, not getting paid was a much bigger problem than corruption.

Doug had much more experience in working overseas and wanted to have the India office as a marketing tool when going after other jobs worldwide. It was good to show we had an office with several hundred people with low labor rates, making our costs competitive and making us look like we knew what we were doing. He was right about this, and we had just developed experience with large jobs that made us a credible international firm.

I didn't argue with Doug but felt very uncomfortable in the situation. After all Tetra Tech was a publicly traded company, and getting dragged through a corruption scandal was not good for anyone, to say nothing of the stock price. I should have just left and retired, but I wanted to finish the work in Odisha and Amritsar. We had also gotten two more big jobs along the Brahmaputra River in the State of Assam, funded by the Asian Development Bank. We were becoming big time international players and that part was fun. Our work was going to improve people's lives and that's what it's all about.

Administratively, I did the cowardly thing and wrote a memo to cover my ass, at least in my own mind. The memo said I thought we should sell the company to Satish and keep a small Tetra Tech team to manage the internationally funded work, subcontracting to Satish for much of the day-to-day work. I had spoken to Satish, and he was definitely interested in buying the local business. This would shield Tetra Tech from the monkey business and still expand our experience with large international infrastructure projects. I didn't mention any of the underlying reasons for the recommendation, although they were well known to Doug. I debated sending it to Dan Batrack, but I respected the chain of command and didn't. Doug didn't act on the recommendation, for the reasons noted above, and never sent it to Dan. Maybe if I had sent the memo to Dan, things would have played out differently. He had a much lower threshold for risk than Doug.

Finally, on the business side, I should give India credit for it's simple tax system. I filed taxes in India in 2010 and 2011, because I was in-country for more then 6 months each of those years. Price Waterhouse prepared them for me, at Tetra Tech's expense. I also had to file US taxes, taking a deduction for the amount I paid in India. My tax filing in India was 10-15 pages. My US filing was almost 300 pages. It was clear then that the US tax system was broken, and it remains so today.

7C. POLITICS.

There are local and national politics, and this was true in India. Local politics tend to influence specific jobs, in the moment, and national politics tend to influence longer term trends in economic areas. Simply put, where is the government spending money? An example of the influence of local politics was illustrated by our 1-million-dollar receivable in Uttar Pradesh, where one administration would not even talk to us and another would open negotiations, even though they were not successful.

Local politics were not complicated once you knew the players and their motive, which was always money. National politics were much more interesting and didn't always end up about money. A case in point was the Ganges River project. On a national level the government was spending money to improve environmental conditions, related to air and water quality. The air quality in Delhi was literally a killer, and one of the reasons I wouldn't live there. The biggest water quality embarrassment was the condition of the Ganges River. It was one of the great rivers of the world, sacred to Hindus, and an open sewer in the dry weather months.

The national government put out a request for proposals to do a comprehensive modeling study of the Ganges river basin and propose recommendations to allocate water for different uses (drinking, agricultural) and to clean it up (sewage and storm water treatment). This was truly a world class project and fit perfectly with Tetra Tech's modeling capabilities and with my experience with river modeling. To be successful, you needed to show strong Indian involvement and not appear to be an international firm full of foreign experts telling the Indian government what to do with their precious Ganges River.

I did two things in addition to having Satish grease the wheels behind the curtain. I found a Tetra Tech employee who was Indian, educated at the Indian Institute of Technology-Delhi (IIT-Delhi), a PhD modeler, and willing to move back to Delhi to be the project manager. Perfect, with me already living in India as principal-in-charge. The second thing, with Satish's help, was to get the chairman of the civil engineering department at IIT-Delhi, who owned a modeling consulting firm on the side, to join our team.

A word about the Indian Institutes of Technology. There are five or six of them in the major cities and they educate the cream of the crop. It's harder to get into them than MIT and they are held in the highest regard nationally, for good reason. Getting one to be a subcontractor was a coup, especially since the chairman of the CE department was a good guy and influential.

So, long story short, we won the project. This lit up the lights back in Pasadena, especially with Tetra Tech's chief technology officer, Leslie Shoemaker. She was also a PhD modeler and wanted to put out a press release right away with great fanfare about this world class project. Fortunately, I asked her to wait until we had a signed contract, since I wanted to negotiate a few details, especially the crazy short schedule, 9 months, for such a monumental project.

We were asked to come to a meeting with the chief minister, who was the equivalent of the U.S. EPA administrator, appointed by the prime minister. We were to give a presentation of our proposal and finalize negotiations. Satish and I went with our PM, who we flew in from California, and the chairman of the CE department at IIT- Delhi. When the chief minister made his entrance, we realized that he was not the one we had been working with, since the prime minister had just shaken up his cabinet. This was a new guy, who went to IIT-Chennai in the southern part of India, and Stanford for a PhD. He was clearly impressive, but I knew from the beginning of the meeting that things were going to be difficult because he was not part of the process that selected us.

All four of us spoke and told the new minister how brilliant our team was, but he wasn't buying our story. Once we finished, he got right to the point. He didn't like the idea that an American firm was the face of the project, even though we had an Indian PM. He also wanted more participation from the IITs, even though we had IIT-Delhi as a key player. He, of course, specifically wanted IIT-Chennai to play a role.

I tried to play up IIT-Delhi's role, but he was not convinced. He finally said we needed to go back to the drawing board and figure out how to involve several other IITs in a meaningful way.

This was a disaster, and I felt he was telling us to get lost, without telling us. We did make a revised proposal, but it didn't go anywhere. We finally heard that he was asking the IITs to come up with a plan to do the work without us. It was bad enough having one academic institution involved, but having them lead the project as a group was a sure sign of delay and failure.

Although Tetra Tech was excited about the project, I'm sure when I told them the bad news, they thought it was another reason not to work in India, and they were right. In hindsight, although I was upset, the expectations were way too high given the budget and schedule, so maybe it was best we didn't go down that rabbit hole. I never heard that the IITs ever completed any part of the work, and the Ganges is still a mess. The lesson is not to count on anything in India until it is done and the final check has been deposited and cleared.

On a broader national scale, it was an exciting time for me to be in India. The prime minister, Dr. Manmohan Singh (2004-14), was an economist and brought India out of its socialist ways into a more free enterprise system. Everyone wasn't working for the government, and capitalism started to work its magic. It was hard for the old timers to change, but the younger folks, and India has more than any other country, loved the challenge. As I have said before, the Indian Tiger will give the Chinese Bear a run for its money over the next 25 to 50 years.

The current prime minister, Narendra Modi (2014-present), is not, in my opinion, good for the country. He is turning India in the wrong direction, not unlike the issues here in the U.S. Hindu nationalism, like U.S. white supremacy, is a corrosive, inward looking policy that stifles growth and hope in the future. It plays on hate, not love, exclusion not inclusion, and is a dead-end strategy. There are approximately 200 million Muslims in India, and Modi wants to marginalize them, or better yet, send them to Pakistan.

I still keep in contact with folks in India, and some think Modi is great. It illustrates the great breach between the Hindu and Muslim religions. The centuries-old problem. I think the Hindus still resent that the Taj Mahal was built by a Mughal (muslin) Emperor. What do we call the period between the 5th and 10th centuries, when the church was the ruling influence in Europe? The Dark Ages. Modi and Trump likely think they were the good old days. We don't want to repeat them.

INDIA BLOG

The following blog captures my time in India from Jan 2010 to Oct 2012, a total of 88 entries. It also includes several short adventures elsewhere, including in Nepal, South Africa, and Qatar. While I was in India the name of the state I lived in was changed from Orissa to Odisha. I use both spellings. Name changes were common throughout India to distance the country from British influence. Enjoy the trip through a fascinating, frustrating, chaotic and high energy country and its people.

FRIDAY, JANUARY 29, 2010

A few Facts

Four times the population of the U.S. One-third of the U.S. land area. 25% of the world's population under 25 years old. More cell phones than any other country. Five of the richest people in the world, and 500 million people in poverty.A land of huge contrast and great expectations.

Rick in his office

SATURDAY, JANUARY 30, 2010

Why India?

There has been a lot said about the economic opportunities in the BRIC countries (Brazil, Russia, India, and China; and sometimes South Africa). These countries are contrasted, favorably in many cases, to the more mature economies of the U.S., Japan, and the European Union, a contrast that held up through the economic meltdown over the last 18 months. All of the BRIC countries were less impacted, relatively speaking. So if you buy the BRIC argument, why pick India? Three reasons: size, democracy, and language. By way of proof, the middle class in India is bigger than the population of the U.S. Size matters. The country is a messy, frustrating democracy, but it is a democracy and one that is evolving from its early socialist roots to something more user friendly to the private sector. If only we can make it through the transition. Lastly, the official language is English. Every educated person speaks it and there are many, many educated young people who speak and write English better than many young Americans. For someone like me who couldn't learn a foreign language on a bet, English is a BIG deal. So that's why! And the people are wonderful gentle souls socially, and a bit less so, businesswise. My prediction is the Indian Tiger will be eating the Chinese Bear's lunch before too long.

MONDAY, FEBRUARY 1, 2010

The India Operations

I'm here to help grow Tetra Tech's India operation and manage two large projects. First, a word about the India operation. Tetra Tech India Limited (TTI) is a typical engineering consulting operation with over 150 people, mostly in Delhi, doing civil infrastructure and MEP (mechanical, electrical, plumbing) work. The civil infrastructure work focuses on the water sector, consistent with Tetra Tech's overall strength. The MEP work has been developed by Wardrop (a Tetra Tech company) and focuses on the mining sector, with a keen interest in the nuclear sector

Satish Malik is the managing director of the India operations. He is a chemical engineer, but we won't hold that against him: I'm a civil engineer. The good news is that Satish has an entrepreneurial gene that is so important in growing a business. In short, he is not happy with the status quo and is always looking for the next opportunity. He is also a great guy to work with, has a wonderful wife of 17 years, and two great kids. Satish has made me feel like part of his family, shown below at the Taj Mahal.

Rick with Satish's family at Taj Mahal

There are many other folks at Tetra Tech India who have helped me fit in and they will pop up later, but one is especially worth mentioning now. Poonam Kaul is in charge of HR and constantly helps me navigate the local scene. She arranges things for me, mostly related to travel and administration. She is literally there day and night when I miss a flight, need a cab, a phone number, or just a reassuring word. Poonam is my lifeline and security blanket.

TUESDAY, FEBRUARY 2, 2010

The Projects

The two large projects I'm involved in are a wastewater/drainage project in two cities in the state of Orissa (Bhubaneshwar and Cuttack), and a River Basin Plan for the Ganga (Ganges) River.

The two cities in Orissa have a combined population of 1,000,000, with some sewers and no treatment. The population will triple in the next 30 years, so it is not a moment too soon to improve the wastewater and drainage infrastructure. There are about 25 people working on the Orissa project now and it will grow to 35 in a few months. The first year will include design, and the next 4 years will see construction. The picture below shows the opening of the project office on January 1, 2010.

Office opening, January 2010

The Ganga River Basin covers about 1/3 of India, with a population of over 300,000,000 people. Yes, that's the right number of zeros! Think of the Mississippi River Basin with all the U.S. population crammed in and it's still doesn't capture the Ganga. Its human, commercial, and cultural importance to India is hard to underestimate. We are in the final stages of negotiating the contract with the government of India, so keep your fingers crossed. Needless to say, this is an exciting, once in a lifetime opportunity for an old environmental engineer like me.

WEDNESDAY, FEBRUARY 3, 2010

My Home Away From Home

I'm living in Bhubaneswar, but will spend time in Delhi. Bhubaneswar is the capital of the state of Orissa. There is an old part of town and a new part. Our office and my flat are in the new part of town, which was planned in the 1950s and has wider roads and less congestion. It's a lot less stressful than living in Delhi. Think of the difference between Boston (Bhubaneswar) and New York City (Delhi). Only add a lot more air pollution and about 5 million more people (and cars) to NYC and you can imagine why Bhubaneswar is preferred over Delhi, even if I can't pronounce it.

Don't think I'm out in the sticks. There was a terrific New Year's Eve celebration at the Bhubaneswar Club. A band, as good as any in the States, wide screens, not quite as good as Patriot's stadium, a fashion show, fireworks, and plenty of spirits to get in the mood. They do have rules, but keep in mind that there are endless rules in India, but no one pays attention to them.

FRIDAY, FEBRUARY 5, 2010

Some Random Observations

Two completely disconnected observations about bartenders and barbers. India needs an army of competent bartenders. I spent two weeks trying to get a satisfactory martini, with little success. If I didn't know better, I would suspect this was the reason most Americans go home. Alas, I have continued the quest with poor results. The good news is the local beer,

Kingfisher. It is as good as any in the States and is now my standard drink. It also avoids ice cubes, which if made from tap water can be problematic. The local red wine isn't bad either, especially for someone like me who can't tell a Chateau Lafitte Rothschild from the backyard grapes. But beyond the beer and wine, transporting some bartenders over here, who know how to make mixed drinks, would be welcomed. Tell your bartender friends.

Barbers, on the other hand, are excellent. For the equivalent of 4 bucks, you get the best haircut going. It has always upset me to pay almost 40 bucks for a haircut back home that, in my case, takes a few minutes. To top it off, my haircut here comes with a head massage. Don't tell this to Frank my barber on Franklin Street, who wouldn't be caught dead giving a head massage. Come to think of it, that is fine with me.

MONDAY, FEBRUARY 8, 2010

North India

I just got back from Kashmir, the northern most state in India. We have a project in the town of Leh and met with the chief minister (equivalent to a state governor in the U.S.) to explain our master planning efforts. The plan includes about 80 million dollars' worth of road, water, sewer, drainage, and solid waste projects. It was a very formal meeting with the CM, His Excellency, who is a wise and impressive gentleman with a great sense of humor.

Leh is in the Himalayas at an altitude of 10,000 feet, the highest inhabitable town in the world accessible by road. Its population is about 40,000, estimated to go to 200,000 in 2040. It is a big tourist destination because of the mountains, lakes, and monasteries, which are about 100 miles from the Chinese border.

Unfortunately, we met in Jammu, the winter capital of Kashmir, and I didn't get to see Leh. Maybe when the weather gets better, I'll do a field trip with my fishing pole. Things were quite fine in Jammu. We stayed in an old winter "cottage" of a Maharajah, overlooking a river valley behind Satish in the photo below.

Satish on deck

TUESDAY, FEBRUARY 9, 2010

Kashmir

Some of you may be wondering about my visit to Jammu in Kashmir, given the unrest in the region. In 1947, India was partitioned into India, mostly Hindu, and East and West Pakistan, mostly Muslim. Kashmir, which is largely Muslim, went to India, much to the displeasure of Pakistan. Since then, there has been constant unrest...think Northern Ireland and the Middle East. East Pakistan is now Bangladesh and West Pakistan is now just Pakistan.

The local papers have something about Pakistan every day, often focused on Kashmir. Two weeks ago the Indian police killed (by accident?) a Muslim student protester, which fueled more and larger protests.

I was in Jammu in the southern portion of the region which is largely Hindu and, therefore, the political situation is remarkably dull and boring. There are very strong feelings between India and Pakistan, and they are not good. It might be less of an issue worldwide, except for the fact that both countries have nuclear weapons.

For history buffs, the book "Freedom at Midnight" is an excellent account of the end of the British rule and the beginnings of India and Pakistan. I recommend it.

WEDNESDAY, FEBRUARY 10, 2010

Cricket

I just saw my first cricket match. It was between the Bhubaneswar Club and the Cuttack Club in a stadium that seats 50,000 people in Cuttack. These clubs represent the two largest cities in the state of Orissa. As in the U.S., the teams consist of weekend warriors who are enthusiastic but carry a few extra pounds (photos below). Since this was a local affair with mostly family and friends in attendance, there were plenty of empty seats. I hope to see an international match with Australia next month when the stadium will be overflowing. Cricket is taken VERY seriously here, especially when the games are with Pakistan.

Once you get the hang of cricket, it's much like baseball. In addition to the obvious bat and ball, the pace, strategy, and traditions are similar. One tradition I particularly liked is the beer, and you don't get cut off in the late "innings". The games can last 5 hours or 5 days, likely one of the reasons beer is important. Fortunately we were at the 5 hour version, which started at 3 pm and ended at 10 pm under the lights.

Each team has 11 players, but only 10 bat. The 11th is the bowler, or pitcher. The team that wins the coin toss bats around, there is a break, and then the other team bats. Each batter is up until he makes an out. So in this respect it's not like baseball since the score doesn't go back and forth as the 9 innings progress. On Sunday, Bhubaneswar batted first and took 4 hours to score 146 runs. Since a typical score is between 150 and 200 runs, there was much excitement on the Cuttack side as they came to bat. Sadly it was not to be. Cuttack could only muster 136 runs and lost. I tried to be impartial, but since I have an application pending to join the Bhubaneswar Club, I was rooting for them.

The boys ready to play

Bottoms up

SUNDAY, FEBRUARY 14, 2010

Traffic

Some things are the same all over the world. Traffic isn't, and in India it's not because they drive on the left side of the road. The traffic in Delhi is probably the worst because of the sheer number and type of vehicles, but most of India isn't much better. The main problem is the variety of bikes, scooters, motorcycles, cars, buses, trucks, and of course pedestrians and animals. Everything goes at different speeds in any lane and when you are going over 40 kph (25 mph), you need the eyes and instincts of a fighter pilot to be safe. Add to this little or no driver education, no rules anybody cares about, and police who are casual observers. Things can get interesting.

The good news is I don't drive, so I was able to take the pictures below including the one of my trusty driver, Mantu. All the traffic causes air and noise pollution. Tata Motors has come out with the Nano, which is about $2,000 and has excellent gas mileage that helps with the air pollution. In Delhi and other cities, most of the buses run on natural gas, but with 1,000 cars added to the roads every day in Delhi alone, it's an uphill battle. The noise, or horn, pollution is another story. In the U.S. I don't use the horn and don't know where it's located. In India the horn is as important as the gas pedal. The horns go constantly, to get vehicles or people to move out of your way, to alert them when passing (see sign on bus below) and, I think, just as a reflex when things get quiet.

This is an interesting subject and Tetra Tech India does traffic engineering, so we are looking to expand this market.

Downtown Cuttack, with cow

Modern Road in Bhubaneswar

Blow horn when passing

Mantu, my driver

SUNDAY, FEBRUARY 21, 2010

Kickoff Meeting

We had the kickoff meeting for the Orissa wastewater and drainage project on Friday. It was attended by about 150 state and local officials in Cuttack, including the mayor. Most of the work will be in Cuttack, solving their combined sewer/drainage problems. Cuttack has about 500,000 people, the size of Boston, but because the density is about 10 times greater than Boston, the city is much smaller in size.

There are essentially no sewers in Cuttack and almost no storm drains as we know them. As the pictures below show, the sewage and storm runoff are collected in open trenches in the streets. These discharge into progressively larger open channels which find their way to the rivers that surround the city, which is an island. This combined sewage would be bad enough, except that the city is shaped like a bowl with a dike all around protecting it from the high river waters during the monsoon season. This keeps the river water out, but also keeps the combined sewage in. As the rainwater (and sewage) in the city backs up because it can't flow out against the higher river elevations, much of the city floods. It's a mess and a major public health hazard. Needless to say, the city is anxious (and so are we) to fix this problem with a sewer system that separates and treats the sewage. We will also build an improved drainage system with pumps to lift the water out of the bowl (that is the city) into the surrounding monsoon swollen rivers.

The last picture shows the kickoff meeting where the politicians were supportive, but of course we haven't done much yet to get in trouble. India, being a practicing democracy, has many local folks who will watch us closely and not be shy about speaking out if they are not happy. Sound familiar? They also file lawsuits at the drop of a hat. No one goes to jail but it does tend to slow things up.

Wish us luck.

Small Street Sewer/Drain

Mid-Size Sewer/Drain

Large Main Drain to River

Dr. Panda's (our client) opening remarks

MONDAY, FEBRUARY 22, 2010

More on Traffic

In the U.S., U-turns are basically extended left turns. In India, thanks to the British, they are extended right turns. That's not the problem. The problem is making them on a 4-lane divided road with narrow medians, which are typically fences or landscaped islands. There are two critical stages in the turn. First, as you slow to enter the turn you can't find a safe haven in the median, so your "ass end" hangs out in the fast lane and becomes an increasingly bigger target for the traffic behind you (first photo). This has already resulted in an accident for Rick Wilson and me, which fortunately caused much more damage to the car than the passengers.

The second stage is merging into oncoming traffic, and there is ALWAYS oncoming traffic. The basic approach is to ignore anything that is smaller than you and play chicken with the rest. You follow the only local rule: whomever gets there first has the right of way.

On my way to work, there are two U-turns. Because of the reasons noted above, these are to be avoided. The first one can't be, but the second one can. It occurs at a grade separated intersection of two 4-lane roads with medians, very close to the office. As we approach the intersection, the normal route would be to take the off-ramp (left turn), go a ¼ mile, make the dreaded U-turn, and go back about 1/8 mile to the office. My driver, being a creative lad, chooses not to take the off-ramp, but rather continues under the overpass and takes the on-ramp... in the wrong direction (second photo). Just as we would need to merge into oncoming traffic on the main road, we can scoot off the ramp to a frontage road and the office. It saves a little time, but it's like going from the frying pan into the fire.

I should add for context that going the wrong way, at least for short distances, is quite common in India, and the cows do it all the time (last photo).

Entering the U-turn

Going Off' on the On ramp

Cow taking U-turn

SATURDAY, FEBRUARY 27, 2010

Shopping

There are several ways to shop in India. I'm talking about how the locals shop, not the tourists in fancy hotels. There are a few small enclosed malls in most cities and in the big cities, Delhi and Mumbai, there are huge ones just like Natick Mall and South Shore Plaza. This is a small part of the retail trade catering to the upper middle class and above. The rest of the folks shop with millions of individual merchants. Most sell one thing (milk, shoes, paint, or a particular type of sweets) and the shops are typically clustered by type. So if you want shoes, you know to go to a certain block on a certain street. In the urban areas the streets are lined with shop after shop. Some can be pretty rundown. In Bhubaneswar, there is an ongoing public/private partnership (PPP) effort to upgrade conditions. The city builds the shops (really stalls) to a certain standard and rents them to the merchants. The city also maintains the area between the road and the shops. This is a sore point with the merchants because they claim that the attempt to beautify the area in front of their businesses is taking away parking. Sound familiar?

Cookie store, my favorite

Typical road side shop, but not my tailor

In terms of food, there are very few "supermarkets" as we know them. That is not to say there isn't plenty of food. It just tends to be local, fresh, and sold in bulk. Very little is prepackaged for a long shelf life, so although it's fresh and healthy, you need to be careful that it is fresh enough. In fact, one of the biggest national problems is the poor distribution system from farm to market, especially in cities where 30% of produce is lost due to inefficiencies. It goes bad in the slow supply chain.

Mantu, my driver, and I call my favorite place the cookie store. It, of course, sells cookies, pastries, lunch stuff, and best of all, ice cream. It's right down the street from our office and I have become a regular.

Mini mall, closed in the morning

PPP food shops

THURSDAY, MARCH 4, 2010

Top of the World, Nepal

I spent the last few days in Nepal getting acquainted with the Himalayas. My friend, Peter Swartz, and I got lucky with the weather and had the opportunity to snap several pictures as shown above and below. It's hard to describe, so just look, and if you think it's better in person you would be right.

Like India, the people are gracious and have many similarities, 60% being Hindu and 40% Buddhist. If anything strikes you as different from India it's the bright colored buildings, giving a gingerbread house feeling.

The two main Cities, Kathmandu and Pokhara, are about 3,000 ft. in elevation and the mountains are in the 20,000 to 29,000 ft. range. Neither city gets snow and the weather in March is summerlike in the day and cooler at night. In short, perfect. You pass banana and mango trees along the roads while looking up at the mountains above.

The hilly land below the mountains, where the cities are, is crisscrossed with rivers draining the snow melt and monsoon rains. The terrace farms on the hill sides are designed to prevent erosion, but deep 1,000-ft. gorges are dramatic reminders of the power of moving water. All in all an amazing place.

Gorge with Mountains in background

Colorful House in Kathmandu

Mt. Everest, center left peak

MONDAY, MARCH 8, 2010

More from the Top

Despite all the beauty, Nepal is a poor country, poorly run and stuck between China and India. They like the Indians more than the Chinese, but not much. Most of the talented people leave. We had our trip arranged by Dr. Sunil, a colleague from Kathmandu who is working in India. His younger brother, Nabin, was our local host, but he too has applications pending in schools in the U.S. and China.

Because of the topography, there are many isolated old kingdoms centered around a public square. There are assorted temples supported by surrounding agricultural land, much of it terraced on steep hillsides. It all make you want to stay awhile, drink some tea, relax, and marvel at the eagles soaring overhead, past the mountains.

Happy Family

Busy Town Square

SUNDAY, MARCH 14, 2010

Holi

Holi, like many religious holidays in India, has gotten away from its roots as a celebration of the coming of spring. Like most things in India, has multiple interpretations linked to many of the thousands of gods worshiped in Hinduism. The modern version, as the pictures below show, is a wild colorful affair, where old clothes are a must. You greet everyone with a shower of colored powder and water and a big hug. It's an up close and personal mess that makes for quite a party. It starts in the late morning and goes all day with lots of food and drink.

The modern version also uses powder that can be washed off with a little extra effort, unlike the older powders that took weeks to fade away.

Getting started with Satish's kids

In the thick of things with Satish's friends

MONDAY, MARCH 15, 2010

Statues

We found a statue maker in Orissa that was a great surprise. The art is an old one handed down generation to generation. The first picture below shows the first step in making a statue out of stone. There are several intermediate steps and each one is done by a different person, so trained. The shop owner's specialty is women's faces.

Most of the larger pieces are made of sandstone and the smaller, more valuable ones are made from harder stones of darker color. They also do wood carvings, but in much smaller numbers. There is a gallery with the nicest pieces, but the power was out so the pictures didn't come out in the dark. Stuff you would expect to see on Park Avenue or Newbury Street.

There are Hindu and Buddhist statues. The big Buddha in the last picture took 7 years to make and is about 10 feet high. After all that, the person who commissioned it backed out. You can have it for your own backyard for about $100,000 U.S. Double the cost for shipping.

The raw material

The finished product

The Big Buddha

WEDNESDAY, MARCH 24, 2010

Chilika Lake

Chilika Lake is Asia's largest lagoon. It covers 600 sq. km. (150,000 acres) at low water, and doubles in size during the monsoon. It's located only 1.5 hours south of Bhubaneswar and we were there during dry weather. It's huge and visible from space even at low water. Think on the scale of the Everglades. It's claim to fame is a stopover for birds that migrate from northern Russia to miss the winter. We saw lots of birds, but by the time we got there in March, most of the big flocks had left heading north for the summer. The other attraction, the dolphins, were still around. We saw a bunch but couldn't get a good picture.

There is a lot of fish farming for prawns (shrimp) and the villages have plenty of local fish that is quite good, reportedly because of the brackish water. I tried to fish, but didn't have the right equipment and wasn't very successful. So we bought some fish and it was indeed good, certainly fresh!

The 'lake' is formed by a long barrier island that separates it from the Bay of Bengal. The island is narrow, 1,000 feet in places, deserted, and you can walk across easily in the shade on a bed of pine needles. As you start to walk from the lakeside you begin to hear the surf and because there is no one around, when you see the ocean you feel like Robinson Crusoe. People come and stay overnight in one of the few primitive shelters, or in tents. You could also sleep under the stars, which offer quite a show on a clear night. One of those great places that isn't easy to get to.

The pictures show a lake taxi, and the beach on the barrier island.

Getting around on the lake

Peter and I on the beach

WEDNESDAY, MARCH 31, 2010

Good and Not So Good

In any new situation, there are invariably little things that surprise. Coming from the U.S. to almost any other country, the not so good typically outnumber the good. India is no exception. Let me list a few in no particular order.

NOT SO GOOD:

• **Power.** In terms of an infrastructure gap, power is the biggest in India. We have good power in Orissa, but it still typically goes out for short periods every day. In other parts of the country, it's much worse. Emergency generators solve the problem, but not always.

• **No clocks.** This is more prevalent outside the big cities and is due to the unreliable power. If there were clocks, you would forever be resetting the time. Watches and cell phones solve the problem.

• **Toilet bowls.** This is also mostly a problem outside of the urban areas. The problem is there aren't any, just a hole in the floor. I laughed so hard one night when Satish was trying to explain the technique. He hadn't used one since his childhood, but the memories were still fresh. After Peter's trip to India, he too can tell a similar story.

• **Kitchen dish racks.** You can't buy one, so what do you do with the wet dishes? Not a big deal since I don't do much cooking, but something that drives me crazy.

• **Cleaning.** There are cleaning products and sponges, but they are not typically used by the cleaning staff. If you want something really cleaned, you need to demonstrate the technique and watch it done once or the results are disappointing. Also workmen, delivery people, and houseboys always seem to have dirty hands, and for some reason they love to touch the walls. I have the cleaning people trained to search out and destroy dirty fingerprints, but it's been a battle, especially in the office.

• **Cleaning II.** I think there are only 3 vacuum cleaners in the entire country and I don't have one. It would be so much easier cleaning rugs if one was available. But since I don't do the cleaning, why should I care.

GOOD:

• **Ceiling fans.** They are everywhere and much needed in the heat. They keep the air moving and with good ventilation, it's sometimes better than AC (and of course takes less energy).

• **Showers.** The water is hot, the pressure is good, and there are no shower stalls, except in the hotels for foreigners. The entire bathroom serves as the shower area, at least the portion where the shower head is located. It's great not to be closed in as long as the floor is adequately sloped to remove water quickly.

• **Holidays.** India is a great country for holidays. There are somewhere between 15 to 20 per year. They make up for the 6-day week. You work 6 days, except for the second week of the month when you don't work Saturday, and a number of other exceptions. To give you an idea of the extent to which holidays are celebrated, Good Friday is a holiday. In a country that is 85% Hindu, this is amusing to me. People think I should be thrilled about this, but as a non-practicing Christian it doesn't move me. We work every holiday and most Sundays anyway!

• **Religion and food.** These two even out the score, but more on them later.

SATURDAY, APRIL 3, 2010

Medical Treatment

You might think that in India the slightest aliment should send you to the local airline office to book a flight home to see your trusted doctor. Well, practically speaking that can't happen – and it doesn't need to.

Last week, one of the expats on the Orissa job, Joh Kang, woke up one night with a fever. He made it through the night and in the morning, two of our trusted local administrative folks took him to the Apollo hospital around the corner from our office. Apollo is a chain of about 8 private hospitals in India. The one in Orissa just opened 6 months ago.

Joh was seen quickly by a series of 3 doctors with increasing degrees of seniority. He was examined, given medication, and they took blood because Joh was concerned about malaria even though the doctors thought the possibility remote. He was back to work the next day and the malaria test was negative as anticipated. All in all the visit and results were most appreciated by Joh and the cost was less than $40 dollars. That said, you need to pick your hospital well. There are good and bad private hospitals and the public ones aren't up to the standards of the private ones. As the two pictures show, it's not too hard to pick the good ones.

Apollo Hospital. The good one.

The other one

THURSDAY, APRIL 8, 2010

Education

Yesterday, the U.S. Ambassador to India, Tim Roemer, came to Orissa. The main reason was to visit an extraordinary school in Bhubaneswar, actually two schools. One, Kiit University, is a conventional 10,000 student operation that includes engineering, social sciences, and a medical and business schools. The other, the main focus of the trip, is a K to 12 school that teaches 5,000 kids from the tribal areas in Orissa for FREE, including room and board seven days a week. The founder of both schools is quite a guy and I had a chance to spend some time with him. He started in 1994 with 125 kids and, as they say, the rest is history.

I took the photos with my camera, so they aren't great. You really had to see 5,000 kids in uniform and white hats and hear them chant as the Ambassador came to the podium to speak. To say this is God's work is an understatement, and being the only American except for the Ambassador's staff, I had a VIP seat at a remarkable event.

Some of the 5,000 kids, teachers in front

The news conference.
The Ambassador is the tallest one in the middle.

FRIDAY, APRIL 9, 2010

Crocodiles

Peter and I took an overnight trip to the Gahirmatha Wildlife Preserve to try to see the estuarine (salt water) crocodile, the largest in the world. Unfortunately the best time to see them is in the winter (December to January), when the water is cooler. They like to lie on the mud flats and bask in the mid-day sun. I wanted to go in January, but the preserve was closed to count the critters, or so we were told. It turns out that one of the visitors had gotten in a small canoe and a croc grabbed his arm and dragged him to the bottom. A body was not recovered. This, of course, added a bit more excitement to the trip and we decided not to bring our swimming trunks.

There are lots of deer in the area and when we saw one close to the water, there was a pair of eyes off-shore keeping watch. We were hoping for a "Wild Kingdom" moment, but nothing ever happened. One of the local guys said the crocs pick off deer regularly. They also have a sweet tooth for the local villagers. Walls, a few feet high provide their only defense. The locals would wave to us, from the wrong side of the walls.

We saw lots of small crocs, some 8 to 10 footers and one big one. He/she was under some overhanging branches, about 15 feet long. The really big ones are over 25 feet and except for a few eyes, we didn't see them out of the water.

The picture is the path to our cabin, a long way from the water and out of croc range!

Little Crocodile

The path to our cabin with monkey

MONDAY, APRIL 12, 2010

Observations

A few things you come to know after spending some time in India:

1. Time. When you are told something will take 15 minutes, plan on at least half an hour, but to be safe 1 hour. Get the idea?

2. Snacks. At a restaurant, snacks mean appetizers. There are normally 4 or 5 selections for the whole table which cover veg and non-veg foods. This is more than a meal, so you need to pace yourself. Once I figured this out, I'm in the habit of just eating snacks and not ordering a main course. However, the first time I thought the snacks **were** the main course, so I had two meals. It's the same if you go to someone's home for dinner. It's snacks for several hours and then dinner at 10 to 10:30. Makes for a long night.

3. Paperweights. In the U.S., paperweights are mostly for decoration. In India there are fans going in almost every office, certainly the ones with poor or no AC. Much of the bureaucratic paper is still paper, so to keep the papers from blowing all over the place and to hold down the top left corner of a paper you are writing on, a paperweight is very useful. And also a decoration.

4. Heat. Speaking of AC, it was 44 degrees C today. That's 111 degrees F, and the worst is yet to come in May and June. As Kipling said, India is about Heat. You couldn't work in the mid-day without AC. In Orissa , which unlike Delhi is close to the sea, there is a nighttime breeze that cools things off and makes it bearable if you don't move around too much.

5. Yes and no. My favorite is the India habit of shaking your head. We are familiar with up and down meaning 'yes' and sideways meaning 'no.' Indians mostly shake their heads in a rolling sideways motion and it doesn't mean 'no'. It means either 'yes' or something akin to 'I'm not sure'. It takes a while to get used to this because it's so different from what we are familiar with. As you might imagine it's particularly important if you are negotiating something, like a contract. After making an elegant argument about some point in a contract, I would see the roll of the heads and think 'oh shit' when in fact they were (maybe) agreeing with me. Doing business in India can be frustrating, sometimes in the extreme, especially for someone like me who likes black and white. In India, you always seem to be working in the gray area. The good news is that, despite these cultural challenges, human nature is pretty much the same all over the world.

THURSDAY, APRIL 15, 2010

Buildings

Almost all urban buildings are built from concrete or brick, and the vast majority are less than 5 stories. Wood is not used for residential construction because it's a big deal to cut down a tree in India, either on a commercial scale or one at a time. You need a permit and they aren't easy to get. India's contribution to stem global warming.

The big bricks in the second photo below are laterite, cut from open pits. It's a soft, structurally weak stone used for walls and one story buildings, mostly homes, because it's cheap, if available. When a piece of land is subdivided, the owners typically mark the boundaries with these laterite walls. They are everywhere. Most construction is done with concrete, and anything over two stories needs scaffolding and temporary support. All the scaffolds and supports are made of bamboo, as shown in the photo. These can go up 10 stories or more and it looks like a child's play set, but contrary to what OSHA might say, they are quite stable, safe, and easy to install.

The other thing you often see are buildings that are partially finished, with the reinforcing rods sticking up from the last floor, ready for the next one that never came. This is because you pay taxes on the size of the property and not on the building's total square footage. So when you get a building permit, you get as many stories approved as you can and build what your finances allow. It's hard to argue with the approach, but this creates an ugly landscape, something most U.S. planning board members would have a nervous breakdown over. Sadly, you see it all over third world countries.

Bamboo at work

Laterite bricks marking a property boundary

Reinforcing bars waiting for the next floor

SUNDAY, APRIL 18, 2010

Women

The things that stand out most for me, when I think of Indian women, are their dignity and their attention to their appearance. I'm talking about the vast majority of the women on the lower end of the economic scale. When you see these women in the market or on the street and look in their eyes, they don't show much expression, but as Peter pointed out, they have a quiet sense of knowing. Lord knows women in India have seen it all and have adapted and gained strength from their experiences generation after generation. In short, they get it done, whether it's caring for infants without medical support, cooking without running water, or just keeping the family together. Their work is never done, but they always dress in the most beautiful clothing. Whether it's 110 degrees or rainy, or both, the women are remarkably turned out and stay that way all day. A few pictures illustrate the point.

With a rising middle class and a growing economy, things are changing for women with means. Women over 35 generally still follow the traditional role of caring for the home, parents, and most of all the children. Most women view this role as important and embrace it. When couples get together socially, they separate into two groups. The men talk about work and sports and I don't know what the women talk about. I sense that this is changing for the younger generations. Worldly sophistication is on the rise and I suspect the 20-something kids, especially young women, will change the face of India in the next 30 years. As with any change, some good will come and some good will be left behind. Same with the bad. Stay tuned men, this change is coming and it will have global impacts.

Going to the market with the kids

Home with the kids, and Rover

THURSDAY, APRIL 22, 2010

Cricket II

We had our first Tetra Tech intercompany cricket match on Sunday. It was a fiercely fought battle and Tetra Tech won. Actually we aren't ready for league play, but the talent is impressive, especially among the under 35 set. Cricket is truly a national game and loved by all. Everyone wanted to play and enthusiasm ran high. The "boys" that manage the office went out the day before and laid out the field. It looked very official, at least at the start.

Much of cricket is similar to baseball, as I have mentioned, but the main difference is the bowling, or in U.S terms, pitching. The professional ball is similar in size to a baseball, slightly larger, harder, and heavier. We used a safer version that is similar to a heavy tennis ball, but still brought a first aid kit in case. The big difference is that the ball is bounced before the batter, which complicates judging where and when to swing the bat. The bounce can be short, long, left, or right. Hitting it looks easier than it is.

There are basically three ways to score runs. You get one run if you hit the ball and can run from one wicket to the other without the fielders getting the ball to the wicket before you. Think of running to 1st base in baseball. You get a second run if you can run back to the first wicket where you started before the fielders respond. Think of running to 2nd base. The field is round, 90 meters in diameter (about 300 feet), and the bowling and batting are in the middle. So you can hit in any direction. If you hit the ball out of the circle on the ground, you get 4 runs and if you hit it out on a fly (home run), you get 6 runs.

There was much anticipation about how the old boy from America would do. Well the bowlers were easy on me so I hit two home runs. I have a feeling they will not be so easy in the next game.

The Indian Boys of Summer

Batting, protecting the wickets

Bowling, with a running start

TUESDAY, APRIL 27, 2010

Meetings and Paper

I have been to a lot of meetings in India. Meetings:

- Never start on time. A half hour wait for the chair is not uncommon, except when I'm the chair.

- Always have refreshments, usually regular coffee in a small cup half full and if you are lucky, cookies.
- Always have a sign in sheet that you actually have to sign, in case you weren't there.
- Never start before 11 am and can be as late as 7 pm, usually a gap from 1 to 4 pm.
- Cell phones, my biggest problem. People leave their cell phones on (unless I'm the chair) and freely take calls. There can be several conversations going on at once. The cell phone God wins and it drives me crazy.

Bureaucratic India (the large public sector) is big on paper. If it isn't on paper, it doesn't exist. Nothing can be done unless some form of written permission exists to allow it. Documents need multiple approvals (signatures) and everything is actually read by someone, so be careful what you write. As you might imagine, this is not conducive to independent, quick decision making in the public sector -- but that's nothing new.

FRIDAY, APRIL 30, 2010

Religion

This of course is always a touchy subject, but less so in India because Hinduism is so inclusive and non-judgmental. In short, it's not threatening and perhaps that's why it's appealing to me. I'll explain the Gods, sacred texts, and the cow in another entry. It's as important to get the feel of Hinduism. It's all about proper conduct (karma), plenty of reincarnations to get it right, and ultimately enough self-knowledge to become one with Brahman, the eternal, infinite. The great part is there is no "one way". With all the gods and goddesses (there are reportedly 330 million), your path is a matter of personal choice and you have a lot of options. This is best illustrated in that every home has a worship area used every day, not just once a week. This is not to say it's easy and a quick learn. Everything has many layers and if you are willing to take the time and intellectual energy to peel them back, there are great lessons behind the colorful stories about good and evil.

Coming from a Christian background, there is one story that especially intrigued me. By one account, Jesus spent his younger years, that are unaccounted for in the Bible, in India learning from the great teachers of his day. Clearly by Hindu standards, Jesus was well along the reincarnation cycle and turned out to be a fast learner. He left with some good ideas and as they say, the rest is history.

Of the many paths to Brahman, Hindus would say that Christianity, Islam, Buddhism, etc. are equally valid. Remember that Hinduism is the oldest religion, has no founder, and no central authority. Sound pretty good? I must caution, that despite the apparent openness and tolerance, the Hindus in India hate the Muslims in Pakistan. This is due in part to nationalism and religious dogma. In 1947 when India gained independence from Britain, the country was divided into India (Hindu) and East and West Pakistan (Muslim), to avoid violence. It didn't work. A bloodbath occurred as millions of Hindus were driven from Pakistan to India, and Muslims, also in the millions, were driven from India to Pakistan. Another sad example of human intolerance, still felt today, as the U S Civil War still echoes today.

SUNDAY, MAY 2, 2010

Pipes

With the huge water main break that I just heard about in Boston, that impacts millions of people, I suspect the general public will get an engineering lesson in pipe construction over the next few weeks.

We also need to deal with the issue here in India as we plan to build over 400 km (250 miles) of new sewers. They aren't quite the size of the 10 foot water main in Boston, but the principles are much the same. When you are digging trenches deeper that 2 meters (6 feet), and sometimes less, the key problems are high groundwater, sandy soils and bedrock. Unfortunately, we encounter two of the three, high groundwater and sandy soils, where we are working in the coastal plain of eastern India.

The theory is easy: de-water the trench with wells and pumps and shore up the sides so you can work safely. The photos below show a shoring technique developed by a German company that is being used in sandy soil with a high water table in Orissa. Also shown are the well points that help depress the water table to keep the trench dry and, in the second picture, a traditional method of shoring up a trench.

The trick of course is execution in narrow crowded streets with construction crews that aren't always well experienced in dealing with the unexpected, or for that matter the expected. Hopefully, the Boston boys and girls get their pipe work done faster than the folks in the first photo.

I had some construction experience, building a road, the year I was in Vietnam, but I never followed that up with more construction/field experience when I started work with Metcalf & Eddy. This was a mistake as a young engineer, since the more experience you have understanding construction, the better design engineer you will be.

Conventional shoring in India, trench is hand dug

Shoring with German technique and well points

TUESDAY, MAY 4, 2010

Dinner

We have established Sunday dinner as a social event for those of us who have relocated to Orissa from elsewhere in India and around the world. To support the relocated staff, we have 3 guest houses and an apartment for me. One of the guest houses is for foreign expats in Bhubaneswar and two are for local experts, one each in Bhubaneswar and Cuttack. They are in residential areas and have rooms with attached baths and central living and dining areas. The one in Cuttack has the living space on the top two floors and the office on the ground floor. Short commute. They all have air conditioners and internet connections, and each guest house has a cook and staff to support the operation.

This is not easy to set up and Chinmay deserves a lot of credit for making it all happen, although I think it's a talent he is not anxious to use again.

Back to the dinner. We have the best of the cooks prepare the weekly group dinner, since our best cook is quite good. He starts preparing the meal early in the day, sometimes the day before. The actual eating starts with snacks which included fruit and cheese, nuts, shrimp, French fries with ketchup, a local favorite etc. This lasts for hours and finally dinner is served, as shown in the photo. This particular night was mostly veg, some fish, and a little mutton that was lost in its sauce. Salad (tomatoes and cucumbers, no dressing) several vegetables (cauliflower with noodles and cheese, my favorite, corn, carrots), rice and dal, a bean soup, all made for a feast. Add several kinds of bread and top it off with two desserts, pudding with fruit and chocolate cake, another local favorite.

On this particular occasion, the French brought the wine (of course), the Americans brought the beer (of course), and the Japanese brought the fish. We watched the India cricket team beat the Australians during snacks, and dinner started at 10 pm, typical. A good time was had by all with much talk of sports and politics. Everyone slept well that night.

Chief cook standing third from left. A key team resource

SATURDAY, MAY 15, 2010

The Village

I have become friendly with our local agent in Orissa, Kalyan Chhotoroy. Kalyan helps us navigate through issues with the local establishment. The other day he took me to his village. Keep in mind that he has never lived in the village, just spent vacations there when he was younger. His grandfather had a great deal of land and 3 sons. He built a big 3-story house in the village expecting his sons would settle down and tend the land. All three promptly left for the city when they got married. Kalyan was born in Cuttack, about 1.5 hours from the village. His father and mother still live there and run a small manufacturing company that makes silverware.

Part of the reason the sons left was because the government took much of their land in the 1950s and 60s when India was into socialism and the redistribution of wealth, which was largely land in these parts. The country has since seen the light and turned toward the west and a more free enterprise system, although the older generations cling to the past.

The village has three main crops. Rice is the natural one and the fields can be seen in one of the photos. The planting is at the beginning of the monsoon (June/July) and the harvest is in the winter. The second two crops are mangoes and cashews. Both grow on trees that Kalyan's grandfather planted decades ago on barren land. Today the trees are auctioned to tenant farmers who tend and harvest them. It's harvest time now, so we picked and ate some of both. Great snacks and good for you.

Grandfather's house and courtyard

Kalyan, the caretaker, and the dogs

The other side of the tracks in the village

Community well with rice fields behind

Kalyan with the mango trees

FRIDAY, JUNE 4, 2010

Slums

I was back in the U.S. and took a break from the blog. As I return, an overwhelming fact of India, beside the 110+-degree temperatures, reemerges: the vast number of poor people in the urban areas. The attached pictures gives a small sense of what is actually a "high end" slum area in Bhubaneswar. The bigger cities are worse, remember Mumbai (Bombay) in Slumdog Millionaire?

The slums are generally well-identified areas in most cities because they have no public utilities (water, sewer, electricity, certainly not cable). These are communities, not people living in the streets, and many of the people have unskilled jobs. USAID has a pilot project in Bhubaneswar to bring water to several slum areas. It seems to be working, but it's literally a drop in the bucket.

The problem is sheer numbers. People come to the city to get a job and are happy with 5 dollars a day to haul dirt, carry concrete, or split rocks on a construction job. The cities just can't keep up with the volume and the slums pop up and expand. It's the biggest problem in India and no one has an answer.

Slum with corner store

MONDAY, JUNE 7, 2010

Hinduism II

Hinduism is about your conduct and moral behavior and is not as prescriptive as Christianity or Islam. It accepts almost any path you take and as previously noted, has several hundred million deities to fit your personal choice. But there are three prominent gods and goddesses that are manifestations of Brahman, the ultimate reality. They represent creation, preservation, and destruction, and every male god has a female counterpart who is, sometimes, not surprisingly, much more interesting

Brahma represents creation, so he isn't too active now that the universe is well on its way. Saraswati is his wife, the goddess of learning. Vishnu is the preserver of all that is good in the world. His consort is Lakshmi, goddess of wealth. Shiva, the destroyer, is necessary for new creation and is a great action hero. Shiva's consort is Parvati. Each of these gods and goddesses can take on different forms as they come to earth to reveal some truth, right some wrong, or just cause some mischief. These missions and their day to day activities form the basis of endless tales.

One of the more famous tales tells of the elephant-headed Ganesh who was created when Parvati gave birth to a normal boy without Shiva's knowledge. See any parallels in Christianity? Anyway, Shiva, returning from an adventure, wanted to see Parvati who, at the time, was bathing and being protected by her son. As the boy tries to stop Shiva, who he doesn't know is his father, his head is promptly taken off. Once Shiva realized what he had done, he promises to replace the boy's head with the first one he can find. He happens across an elephant, severs the head, replaced it on the body, and thus we have the famous Ganesh, god of good fortune.

The moral is known: know your father.

SUNDAY, JUNE 13, 2010

Sun Temple

We took a trip to Puri about an engineering problem our client is having related to a wastewater discharge to the Bay of Bengal (more on that later). We took a side trip to see the Sun Temple in Konark, about 20 miles up the coast from Puri. Actually the Temple was on the coast when it was built in the 1200s, but is now 2 miles inland. Perhaps global warming will correct this.

The Sun Temple, a World Heritage Site, was built to celebrate a Hindu victory over the Muslims. Others say it was built on the spot where the King's son was cured of leprosy. Like many things in India, there are multiple paths to the truth. Regardless of the reason, the temple was oriented so that the morning sunlight would shine on the feet of the God inside the temple. The temple, which was originally three connected buildings, was constructed as a cosmic chariot with wheels and horses. Very impressive.

One of the brilliant British Viceroys in the 1900s thought it was best to preserve the temple by filling the interior with stone and concrete. So today you can't go inside, but there is much to see outside. Our guide offered us a 1 hour, 8 hour, or 3 day package. We took the one hour tour, since the temperature was over 100 degrees. The guide spent a good deal of our hour explaining the main theme, having babies. And, of course, in order to have babies there needs to be sex, which explains the hundreds of explicit erotic carvings around the temple. The last photo is one example of the wealth of information on the subject, and actually made us momentarily reconsider the 3 day package.

Entrance to Sun Temple

Explicit carving. Lifting the leg improved fertility

FRIDAY, JUNE 18, 2010

Rural Markets

Rural food markets are typically open several days a week, and the vast majority of the produce are fruits and vegetables. This is not surprising in a country that is 70% vegetarian, with a higher percentage (100%?) in the rural areas. In many ways the activity is the same as in Boston's Haymarket, without the meat shops. There are two other big differences. Here you can pick out the fruit and vegetables you like so the usable percentage is close to 100%, unlike Haymarket where you don't get to pick and the spoilage can be huge.

Second are the cows. They know the drill and gather around the outskirts of the market. Each one takes a turn wandering down the crowed lanes bumping into everyone, and when the time is right they grab a fruit or vegetable. They would grab a second except the vendor jumps into action and whacks the cow with a stick to protect his business. The cow scampers away and then another one tries her luck.

We bought two shopping bags of produce for about 200 rupees (4 dollars).

The Market

Setting the price

The cows wait their chance

MONDAY, JUNE 28, 2010

Wedding

Weddings in India are a big deal that last days and involve family, friends, and the whole village. In the U.S., a good size wedding is 150 people; in India, it's 1,500. Chinmay, one of Tetra Tech's rising stars, is from Orissa and is key to our success here. His brother got married in June and half the local office went to one of the receptions. The wedding was in Chinmay's hometown which is about 150 km south of the office. It's right on the train line so we all took the train, which beats the car in time and comfort. If you ride business class, the trains are a great way to travel. It cost about $8 dollars round trip. Weddings start in the early evening (in the northern part of the country) with a visit to the temple for a short ceremony. Than the marriage party and friends form a parade with cars, carts, animals, music, and fireworks. The parades typically cause massive traffic jams, which no one seems to mind. The destination is the bride's home where the real ceremonies begin, which last most of the night.

The next day includes social activities, but at a lower energy level. On the third day, the pace picks up again with a reception for a wider circle of friends and the village. We attended this reception and the picture shows the key players. The bride and groom are beautifully dressed in the picture below, but for the wedding the bride wears red, not white, and the groom dresses in a traditional Indian smock, usually white.

The bride and groom flank the little girl. Chinmay is at far right.

SUNDAY, AUGUST 15, 2010

Monsoon

Much has been written in the last month about the terrible monsoon rains causing flooding in Pakistan and mudslides in China. Except for a few spots in the north, the monsoon has been below normal in India this year. In fact, some states are in a drought. However, there is still about a month to go.

I took the pictures below of the Mahanadi River during the dry season in March and last week. The pictures were taken from the embankment that protects the City of Cuttack from flooding. You can see more water in the recent picture, but it doesn't fully capture the impact of a wet monsoon when the river can be 30 to 40 feet higher than shown. I'm hoping for a week or two of heavy rains so the river rises and we can see how the flood control gates in the embankment work, or don't. I shouldn't really wish for this since the high water can cause flooding both in the river and city. The rain that falls on the city can't escape past the closed flood gates that are keeping the higher river water out. This, of course, floods the city, but what make things really bad is the water is a mixture of rain runoff and sewage. This is the problem that we are trying to fix, so it's always better to see it firsthand then to hear stories. On the other hand, if there isn't bad flooding this year, the public pressure on us to perform will be a bit less intense.

One of the advantages of living in the northeast U.S. is infrequent severe weather conditions. A little basement flooding or beach erosion are child's play compared to a world class natural disaster.

Mahanadi River in the dry season. Highway bridge in background.

Mahanadi River in wet season. Highway bridge in background.

FRIDAY, AUGUST 20, 2010

Mid-Year Meeting

Tetra Tech India Limited, a wholly owned subsidiary of Tetra Tech Inc., had its mid-year meeting at the Siddharth Hotel in Delhi last month. During the day, senior managers and selected middle managers reviewed progress and set the course for the remainder of the years and beyond. See the happy group below. I gave an overview of how the India operation fits into Tetra Tech Inc. and the support that Dan Batrack is giving us as Tetra Tech looks more and more overseas for growth opportunities. Our mission is to grow the local business profitably, in part by winning large donor-funded (JICA, ADB, World Bank) infrastructure projects and to support Tetra Tech projects outside of India with technical talent. As with most business meetings, in addition to discussing markets and technical stuff, there was also a focus on improving systems, controlling costs, and maintaining quality controls.

After the working sessions, the whole company was invited to an evening of entertainment and good food. Indian women and men love a party and it's not only the Bollywood stars that dance up a storm. Everyone does, and I mean everyone. There is a wonderful innocence and lack of worry about making a fool of yourself that is unlike the U.S. I find it refreshing. However, in the Indian culture it's still common for the boys to dance with the boys and girls with girls. I must admit I'm still struggling with this concept, especially when I get asked to dance by another guy!

Before

After

WEDNESDAY, AUGUST 25, 2010

The Bay and Resort

When we took the trip to the wedding, we stayed at a seaside resort on the Bay of Bengal. For those of you without a map, that's the part of the Indian Ocean off the east coast of India. It's actually the Bay in Kipling's **Road to Mandalay**: "when the dawn comes up like thunder out of China cross the **Bay.**" A favorite of mine.

It turned out I was a bit late to see the dawn, but it was overcast anyway. The beach is beautiful, but the surf was rough and we were cautioned not to swim because of the rip tides. There was a safer place to swim, but it was several miles up the beach. We just went in far enough to cool off. It was 100 degrees F at 7 am. The nearest city is about 20 miles inland, and it's small by India standards (a half million) so the water was of good quality (free of fresh sewage). It was a bit turbid due to the surf and sand. There is fishing as you can see by the boat in the picture below. The water is mostly shallow for a mile off shore, so the fish are typically small and caught in nets for local consumption.

Fishing boat on the beach

The resort is behind a huge sand dune that protects it from the sea, but blocks the view. The photo of the bungalows and main building was taken with the Bay to my back. It was the off-season so I had a whole bungalow to myself. It had clean sheets, running water, and a light (when the electricity was on). Always carry a flashlight while traveling. Few creature comforts, but you could hear the sound of the surf and a rip roaring thunderstorm during the night. We got an off season group rate of $10/night that included breakfast. Not bad if you happen to be in the neighborhood.

The resort

WEDNESDAY, SEPTEMBER 1, 2010

Construction Quality

Construction has a serious problem in India: quality. There are many reasons that include lack of qualified contractors, standards-of-care that aren't followed, and corruption. Not only are the main products often sub-standard (e.g. poor concrete; see photo), but the attention to detail is lacking and the last 5 to 10 percent never seems to get done. I'm not talking about the high end 5-star hotels, I'm talking about standard residential and commercial construction. I have complained more than once about regulations and the complicated approval process in the States. There are too many regulations, and the process slows things up. The opposite is true here. The only approval you get is a "Yes," with how many square feet or units you can build. There are many written standard practices, but none are followed or enforced. There are subdivision and environmental regulations and permits are issued with many conditions, but once a project is started, no one follows them. Consequently, things like setbacks, landscaping, sidewalks, curbs, and gutters are ignored. Compounding this, projects are never completed, so that construction debris becomes a permanent part of the landscape. There certainly isn't any architectural review.

There is a clear attempt to change this, and it's working somewhat in the higher-end residential gated communities. These are typically 25 to a few hundred units (apartments or townhouses). They have 24-hour security and the streets are clean and landscaped. However, inside these communities, where I live, the quality problem still invades. For example, my kitchen was painted. The job was quite good, but the painters didn't take the time to put down drop cloths. I have nice white walls, but white paint is spattered over everything else. I really don't think this problem occurred to the painters, who considered the job well done. Changing habits is slow and difficult here, as it is elsewhere.

Concrete construction

THURSDAY, SEPTEMBER 2, 2010

Follow-Up

To further illustrate the point in the last posting about completing jobs and attention to detail, I offer the front of our office building. We have nice office space by any standards. The two pictures below are the reception area and my office, which has a private bath and shower. Very comfortable. Notice the Ganesh statue under my desk. One of our valued senior engineers nicely pointed out that because of where I placed the statue, it could be understood that I was "looking down" at the Ganesh. This was disrespectful and not appropriate, so I moved the Ganesh against the far wall. Lesson learned.

The third picture is the front of our office building. We are on the top, third floor, which you can't see in the photo. In India, our first floor is called the ground floor and our second floor is their first. Not a big deal, but another lesson learned.

The good news is the street was paved. It was dirt at the beginning of the year. I sometimes get delusional and think it was paved for me. Fat chance. So far, one of three white lines has been added. I suspect the other two will come at some time, but the edges will be in perpetual disarray. Construction debris and trash will be our constant companion, and a general state of messiness will persist on both sides of the road. Your personal space (home, office) is generally well-kept. It's the public spaces that go to hell.

This relates to one of my biggest frustrations, the staircase to the third floor. It looks like it's used for finger painting with dirty hands. I requested it be washed, but those thumb prints are difficult to remove. The next step is a paint job, but remember what I said last time about paint jobs. Our nice black marble steps would be forever speckled white. What to do?

Enough complaining. I'm grateful for a nice office and reliable power, which is important in this heat and not always the case in the rest of India.

The reception area with the lovely Malabika at her post

My office, with my good luck Ganesh next to the desk

The front of our office building, messy edges

SATURDAY, SEPTEMBER 11, 2010

Golf

Kaylan recently joined the one members-only golf course in Bhubaneswar and is learning the game. It's gaining popularity in India and just recently an Indian won a major tournament for the first time. We started play early to beat the heat, but that only means the difference between 95 and 100 degrees. Actually, it's the humidity that's the killer, but the grass likes it. With all the rain, even in a dry monsoon, the course is in great shape. I'm not sure what the grass is on the greens, but they were very slow. You had to hammer each putt.

Kalyan has a wonderful set of clubs, but they are right-handed. I play left-handed. He found the only left-handed club in town, a 7 iron which didn't work to well off the tie. I started using his woods and actually did OK right-handed, but that isn't saying much.

They have golf carts, but we chose to walk since there is plenty of help. Since we only had one golf bag, we needed only one caddy plus Kalyan's instructor and a few extra kids to find the balls. I especially liked the extra kids because of my lack of accuracy and the adjacent jungle. There are scary/dangerous critters (snakes and the occasional elephant or tiger) off the fairways, since the course was outside of town and bordering the jungle. The kids didn't seem to mind, and for a few rupees each, we never lost a ball. The course is pretty flat with a huge rise in the middle that includes a large rock out-crop. The clubhouse sits on top of all this and is quite impressive. The people who play golf are clearly upper income and would fit in anywhere in the U.S. We had two guys play through us -- we were slow --and they were very good golfers. I'm going to try again when the heat breaks.

Mr. Otaguro putting. Clubhouse in the background

Me driving right-handed with Kalyan looking on.

The spectators

MONDAY, SEPTEMBER 13, 2010

Cuttack Office

We have a branch office in Cuttack, a city of 1 million about 20 miles west of Bhubaneswar. It was the capital of the State of Orissa until the 1950s and has a 3,000 year history of habitation. It's like the North End, only the streets are half as narrow and the population density is 10 times, with sporadic garbage collection. Fortunately, our office is in the newer part of town, the Cuttack Development Commission District (CDC). We actually rented a combination office and guest House. The ground floor is the office which can accommodate about 15 people, and the first and second floors are the living quarters for the Indian expats. There are 5 bedrooms, a kitchen, and common area for the expats. The cook, who lives onsite, provides three squares a day and the local junior staff sleeps in town.

The facilities are quite nice, the food is good, and it's provided cheaply to the staff. I often find myself in Cuttack at lunch time and look forward to the meal, typically served sometime between 2 and 3 pm. Strictly vegetarian. Unless you are local, there isn't much to do in Cuttack in the evenings, so the staff puts the TV to good use. Last Sunday some of the staff went to a water park just outside of town and reported a good time.

One problem is the birds: mostly pigeons, but many other kinds. In short, they perch in every conceivable spot, sometimes inside the house, and shit all over. This actually happens on most buildings and if I could think up a solution to keep them away, I would be right up there with Gandhi.

First floor, dining room in background

Front of office/guest house

SATURDAY, OCTOBER 9, 2010

Company Dinner

To build a little team spirit, we hosted a dinner for all the staff at the Bhubaneswar Club (more on the Club next time). We planned the event on the roof deck, but since we are not quite out of the monsoon season, we had bad luck and it rained. So, we started out with just another indoor cocktail time and dinner, but Chinmay, always resourceful, suggested we get to know each other better by each giving a little performance. Singing, dancing, a joke, or anything crazy would do. There were a lot of good singers and dancers, I was not one. I did quote a few verses from Rudyard Kipling's "Gunga Din," to a rapt audience. A good time was had by all, even inside.

I'm always intimidated about this, or any kind of group participation. Of course, a few beers help, but it's remarkable how liberating it can be and the power it has to bring a group closer together. Unlike many India functions, which tend to segregate levels of society, we invited everyone to the dinner, including the drivers and cooks -- everyone. It was a wise decision. Not only were the less exalted folks often the funniest doing the skits, but the bond of loyalty that was forged will not be broken.

Part of the local gang, working up to the performances

TUESDAY, OCTOBER 19, 2010

Holidays

As I have mentioned before, there are many holidays in India. Like most places, some are religious (not just Hindu) and some are national. Recently we had back-to-back national and religious holidays. The first was Gandhi's birthday. An obvious one here. The second was for the Ganesh and his father, Lord Shiva. For this one we had a ceremony at the office as the pictures depict.

You get a dot on the middle of your forehead from powder that is blessed by the god, and a string around your arm that connects you to the god and protects you. Good traditions both.

There are also sweets, present at every occasion much to my approval, and the kids. Kids are the same way all over the world. Some are the life of the party and some are petrified. In India there are many more "all family" event than adults only events, especially away from the major cities. The family unit drives the bus here in India.

Rick Wilson with his dot and string being tied

The first kids to arrive are a little tentative

THURSDAY, JANUARY 6, 2011

Flat

I spent more than a month at the Mayfair hotel. Although it's quite beautiful, you start to go crazy in a hotel -- even though I do remember thinking of words like "paradise" when I first checked in. Living in a flat, which has cooking facilities and a cook when needed, is much better than hotel living. I live on the 5th (top) floor in one of 5 buildings, in what is called in the U.S. a gated residential compound, with significant full-time security. The security is standard here, mostly to keep out the masses. I have never felt unsafe in India, and you rarely read about crime in the papers. Given the 1.2 billion people, that's unusual.

My neighbors are upper-middle class. The grounds are kept clean and nicely landscaped and there is a gym and pool, although these are not top of the line, and I have never seen anyone in the pool. There is also a play area for kids. There are many cute kids, and they stare. The apartment is large, with three bedrooms, and lots of windows. The two most important rooms, kitchen and bathroom, take a bit of getting used to. The kitchen is "Indian Standard" with the basic stuff. An iron chef would be horrified, but I'm not one and we have cooks who do just fine. The main problem with the bathroom is lack of storage space and hooks, but that is easily fixed. The other main difference, which I like, is the shower. Actually, there isn't a separate shower, rather the shower head come out from the wall on one side of the room. It gets a little messy, but you never bump your elbow against the shower door. There is a great sense of freedom while showering. One cautionary note. The floors are marble and slippery as hell when wet.

The second picture below is the dining room (my computer is on the table) and the living room. I insisted on wall hangings in the living areas, as I hate blank walls. You can get local prints for next to nothing and have the house boys frame them in no time. The lanterns are also a local favorite, for about $1 dollar each. The wicker furniture is world class, made locally and looks terrific inside or out. I wish I could take it home.

The apartment complex

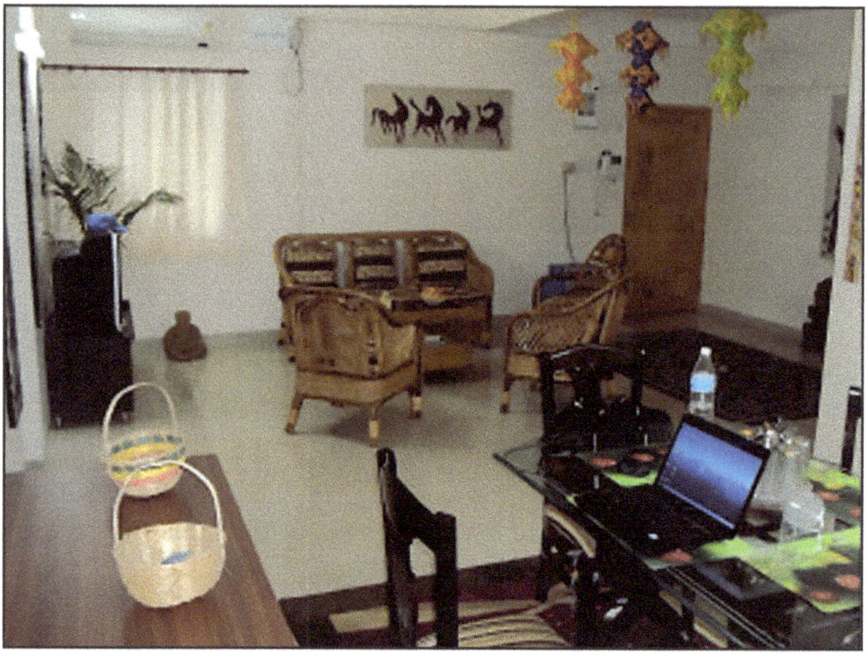

Dining room and living room

SATURDAY, JANUARY 15, 2011

Diwali

One of the main Indian religious holidays is Diwali, the festival of light. It's like the 4th of July, with a little Christmas thrown in. Like the U.S., it's been hugely commercialized with the emphasis on light (fireworks), and a lesser extent on presents. The origin is the burning of butter and oil lamps to lead Lord Rama home from exile. But in practice it's all about fireworks.

Chinmay invited us to his home for the evening. Fortunately, he lives on the top floor of a two-family flat that has a roof deck. We spent between about 7 and 11 pm either on the roof watching other fireworks or setting off our own. There were no tall buildings obstructing the view, so we could see fireworks all over the city that were literally continuous, near and far. Right down the street from Chinmay's flat was a big party with an arsenal of fireworks. Taken in total, it was as impressive at the 4th on the Charles River in Boston, all funded and set off by private families. No reported causalities the next day.

We had some of our own fireworks and managed to set them off without mishap, but it was close a few times. After it was over we had a wonderful dinner and finished off the last bottle of beer.

Our host Chinmay and his wife

Our fireworks and the house decorations

SATURDAY, JANUARY 29, 2011

Street Scenes

If you are out for your morning walk or driving to work, the scenes along the main roads are similar. The poor who live along the road or in slums are starting their day and the many small retail booths are coming alive. The slums have no water or power, so the first thing is to go to the nearest water source, hopefully a community well, and fetch a bucket full. Then breakfast begins over an open fire, with food normally supplied by the government from bulk distribution locations. The more prosperous can buy food at a local market or eat at a greasy spoon and read the paper. No one rushes to get to work in the morning, although they often work late into the evening.

I have never had the pleasure of eating along the road and don't think I will. With all the international folks coming through India these days, I haven't seen many cases of the famous "Delhi belly." If you are careful about not drinking tap water (I don't put ice in my scotch even in nice hotels) and where you eat, you should be OK.

Greasy, very greasy, spoon

Morning vegetables

Slum dwellers

Street Scenes Part 2

The retail folks also start early since people shop almost every day for fresh food. One of the most interesting shops is the chicken stand, although watching one in operation should not be done right before dinner. The live chickens sit in crates and are chopped up all day to meet demand. It's probably easier on the chickens than Purdue, until the end. I eat a lot of chicken, so seeing the details made me a bit uneasy. It's one case where my lousy memory is an advantage. I don't ask our cook where he gets the food, but I hope it's not the place in the picture.

Another interesting shop is the laundry. Fortunately, this isn't Paris and I'm not too worried about the crease in my pants or the amount of starch in my collar.

Finally, there are the street sweepers. Trash and garbage is a **huge** problem in India, as I have mentioned. The cities try to keep certain areas clean, but it's a struggle.

Laundry

Chicken stand

SUNDAY, JANUARY 30, 2011

Republic Day

January 26th is Republic Day in India – the day in 1950, when the Constitution was ratified. It's right up there with the other two national holidays: Independence Day (August 15th) and Gandhi's birthday (October 2nd). There are parades, but by Indian standards it's rather subdued. The tradition is to have a short acknowledgement of the tri-color flag and sing the National Anthem. The other event is a picnic. As with most traditional Indian gatherings, the women and men socialize separately, as the eating arrangements in the picture below illustrate. The Indians are so unself-conscious, and have a great sense of humor, so there is always a lot of laughing and small pranks.

We had our picnic in a park on the side of a hill on the edge of town. On top was a Hindu temple where I paid a visit, gave a small donation, and was awarded a red mark on my forehead for health. It's faintly visible in the picture.

Since we were not the only ones having a picnic on the hill (some groups came by bus), getting back down the narrow road was not the best way to end the day.

Picnic. The guys

Picnic. The ladies_

SATURDAY, FEBRUARY 5, 2011

Mayfair

The Mayfair Lagoon hotel is a patch of paradise that lifts the spirit after a long day of work. I stayed there for over a month before my flat was ready, so I got to know the staff. Actually, I'm not too hard to remember given the number of westerners in these parts.

I go now once or twice a week for breakfast or a light dinner in the British bar. They have two, the one for the younger set is called 10 Downing Street. There are also two nice restaurants for dinner, besides the Tea House for breakfast and lunch. The staff is extremely attentive, which might have something to do with the fact that I tip more like an American than an Indian.

Most of the rooms are around a small lagoon, hence the name, and surrounded by amazing landscaping. The pool is like a movie set and there is an old 1940s prop plane that the late Chief Minister (think state governor) used. He was a pilot and quite a character by all reports. It reminds you of the WWII Flying Tigers, supporting the Chinese and fighting the Japanese.

The whole atmosphere is calming, cool, and clean. It's a great place for a wedding and they have many.

See you. It's cocktail time.

The pool

The lagoon

The plane

Entrance with elephant

SUNDAY, FEBRUARY 6, 2011

Bhubaneswar Club

Besides the Mayfair, I spend time at the Bhubaneswar Club in the evenings for dinner or for one of their events celebrating one sort of holiday or another.

The club was started by the British, no surprise there, and was taken over by senior government officials in the 1950s. It has since been changed to cater to anyone who can get sponsored and pay the dues. This means a range of government and businesspeople on the higher end of the food chain and their families. I wanted a 6-month temporary membership, but was pressured into a 3 year membership for about $1,000. This turned out to be good political form and not too bad when divided by 3. There are several bars, two restaurants, billiards, and snookers (British version of billiards), a swimming pool, tennis on clay, and a gym. They also have several rooms for rent, but not near the standard of the Mayfair in terms of western creature comforts.

The best parts about the Club are the food and drink prices, about 25% of the Mayfair and as good. For this reason, we use the Club to entertain clients. The picture below is of a dinner after a meeting with some clients and Japanese government folks who are investing heavily in India's energy, transportation, and water infrastructure.

In these parts a western Club member is something of a catch, so my sponsor, Kalyan, gets a lot of local points for bringing me in, especially for 3 years. Kalyan is a long-standing, well-known member with the right family background. He is also a competitive snookers player who represented the state in competition when he was younger. Even with my best fraternity form I don't play him for anything more than a 10-rupee (25 cents) glass of beer.

Dinner, Kalyan is the first person on the right looking at the camera

SATURDAY, FEBRUARY 19, 2011

Meeting with the Chief Minister

Periodically we meet with the chief minister about progress on the projects we are doing in the state. The chief minister for the State of Orissa is equivalent to the governor of a state in the U.S. Under the CM are several ministries, no surprise there. Just like the U.S., each ministry is run by an appointed official, like the secretary of transportation in a U.S. state. But the show is really run by the senior bureaucrats, called principal secretaries (PS). It's extremely competitive to obtain this rank, which is based on service and a serious testing process. They move around but typically have a state where they spend most of their time, and the best get appointments from time to time with the central government.

The PS we work with is new. Both the previous one, who went to the central government, and the present one, are top notch people and would be running companies if they were not in public service. It's the many layers of bureaucrats below them that cause the trouble.

All the presentations are given in English, so when I was scheduled to meet the CM the first time I asked if he would understand my English. I was told that he grew up in England and went to Oxford and Harvard. No problem with English!

Chief Ministers Conference Room. I'm on the right (easy to pick out), the PS is two to my right in the jacket and tie, and the CM is to the left, in local dress, at the head of the table.

Local Festival

Every year there is a local festival that lasts several days and highlights the tribal cultures in eastern India. It focuses on tribal art, their way of life, and government propaganda on what is being done to improve things. Parts are like a 4H Club fair where there are exhibits from local villages and farms.

One of the big debates in India is the impact that development is having on the tribal way of life. Since these areas are way off the beaten track, it mostly relates to big mining operations that impact large areas. Some of these lands are sacred or good hunting grounds, to say nothing of rare and endangered species. The Minister of Environment recently stopped a large mining operation for all these reasons. In the past this wouldn't have happened, or some money would change hands and the decision would be reversed. Times are changing and the environment is getting more attention. Progress!?

Local girls weaving

Lanterns in the trees.

SUNDAY, MARCH 13, 2011

Dilip's Retirement

The head of the local Water and Sewer Agency, Dilip Padhi, retired last month. The retirement age is 58, so we may see more of him. We held an event in his honor on the roof terrace of the Bhubaneswar Club. It was a fine affair with cool weather, much food and drink, and the good fortune of having Doug Smith on hand. He was visiting India, so we showed him a little local color.

There were some heartfelt words by those present, a few gifts, some good fun, and Amod sang a song in Dilip's honor. Now Amod had, not long ago, been the lead singer in his own band, so this was not amateur night. I'm sure people stopped on the street below and in the gardens of the Bhubaneswar Club when they heard his beautiful voice coming from the rooftop. I was very impressed.

Despite all the fun, I'm in no rush to my own retirement.

A gift of a picture of Dilip batting at our Cricket game

The Tetra Tech Team: Amod, me, Chinmay, Doug, Darwin, Promod

SUNDAY, MARCH 27, 2011

Rose Garden Part I

To get a bit of exercise in the morning I had been going to the gym at the Bhubaneswar Club, but since I've moved into the Guest House, I have changed my pattern. This is because I'm now only a 10-minute walk to the local arboretum and rose garden. It's a bit of a challenge getting there, but it's worth it. The map below gives you the general plan. There are three main parts: large, landscaped areas with a small lake for boating, the rose garden, and the jungle trail. You walk in one entrance through the landscaped areas with flowers, playgrounds, and trees. The lake is low now in the dry season, so boating is limited. The roses are next before you enter what I call the jungle trail. It's a one lane paved road through the forest that leads to the back entrance/exit. The plan shows a moat around the whole arboretum because there is a large elephant population on the other side. The moat keeps the elephants out. They can be dangerous and occasionally you hear them off in the distance. Really cool, but I'm glad for the moat.

Map of the arboretum

Vic standing on path in the rose garden

Rose Garden Part II

I don't know if you noticed in the last post, but Vic is wearing a Brooklyn Dodger's t-shirt. We were both born in Brooklyn in the same year (not telling which one), but it was long enough ago to have seen the Dodgers play in Ebbits Field and to have soaked up the magic. Vic and I had plenty to talk about during our walks in the park.

We did stop and wonder at the beauty and breadth of the roses. They go on and on and are tended by an army of gardeners. There are so many roses, this must be where they get the petals for the fancy weddings.

The staff for the garden and the rest of the arboretum gather every morning at the maintenance house to get their orders and then fan out to do their tasks. I have never seen a mechanical piece of equipment. Everything is done by hand.

The staff hard at work

A rose by any other name...

SATURDAY, APRIL 2, 2011

World Cup 2010

We normally work on Saturday, but yesterday we had the afternoon off, as did most of the country. India was playing Sri Lanka in the Cricket World Cup final. Like soccer, it's played every 4 years and India last won in 1983. To get to the finals they had to beat Australia, who won the last 3 World Cups, and then Pakistan last Wednesday. In some ways the Pakistan semi-final match was more emotional given the fierce feelings between the two countries, which would need another entry to explain, or more likely a book.

The match got off to a bad start for India. The conventional wisdom is you want to win the coin toss and bat first. In cricket one team bats first and then the second team bats (one inning). The idea is to score as many runs as possible and then let the other team worry about how to beat the score when they bat. India lost the toss.

Your turn at bat ends when all of the 10 batsmen are out (one out per batsmen which is called a wicket) or you reach 300 pitches, whichever comes first. This is one form of cricket that's called a day match. It takes about 8 hours. In this case, from 3:00 to about 11:00 pm. Plenty of time for food and beverage. The original cricket games lasted 5 days, ugh!

Pakistan scored 274 runs. This makes for a competitive game, because the rule of thumb (there are many of them in cricket) is that less than 250 runs can be easily beat, and more than 300 is tough to beat. Everyone was hoping India's first and best batsman, who is a local hero playing in his last game, would score a century. That's 100 runs without getting out, equivalent to 3 home runs in the World Series. Sri Lanka's best pitcher (bowler) got him out in two pitches without a run scored. The stadium, in Mumbai with a hometown crowd, was silent.

Fortunately, the rest of the gang slowly clawed back and passed the 274 mark with just 6 out of 300 pitches left. The stadium, and all of India, were no longer silent.

SUNDAY, MAY 29, 2011

New Minister

The chief minister in the state of Orissa is having a hard time lately with a range of scandals. He is personally respected, but the extent of the beating he is taking in the press is beginning to tarnish his image. Last month he responded by reshuffling his cabinet of ministers. Mostly it was a case of musical chairs, which wasn't judged to solve anything. Worse, the two ministers who lost their jobs have given interviews to the papers blasting him as a puppet, with the strings pulled by sinister forces. Sounds like Massachusetts politics.

The minister in charge of our work was replaced. This required a meeting to brief the new chap on our activities. He is an engineer by training so that is a good sign, but we will see. In the picture below, the minister is at the head of the table, surprise, with the white cover on the chair. The member secretary, next in line, is to his right. During most of our briefing, the minister had his phone connected to an ear plug and spent a good bit of the two hours on other business. I guess the good news is that when he was attentive, he was fully engaged and asked good questions. Afterward we gathered for several cocktails and dinner. Noah, our drainage engineer from California, joined us with his wife who is here with him, and his daughter, who is visiting. They are on the left side of the table at dinner.

Meeting with new minister, on the phone

Dinner to relax

Holi

Celebrating holidays in India is the national pastime. I'm not one to knowingly make a fool of myself, but the good nature of Indians makes it easy and a wonderful time. I'm also consistently amazed at how they have such a good time and consume so little alcohol. This year we had an office celebration at Holi. It was wise not to wear your best clothes, as the photo shows.

Holi was celebrated on a Saturday and Sunday, so the office affair was on Saturday (we work 3 out of 4 Saturdays each month). On Sunday morning, I walked around the neighborhood and the kids were out in full force playing with the colored powder and water. Several of the groups were a bit suspicious. But most just played on, paying no attention to me. A few, however, got me in the game. The picture below is of several kids who had sophisticated equipment to apply the colored powder and water. They had great fun initiating a Westerner into their beloved holiday. I was honored to be part of it all.

The fun

The kids

SATURDAY, JUNE 11, 2011

Medical Stuff

In a previous posting I mentioned the Apollo Hospital, a private chain that just opened their facility in Bhubaneswar. I never expected to see the inside, but some of our U.S. staff had positive experiences with treatment, ranging from minor stomach distress to more serious blood clots. After breaking a tooth on a chicken bone at dinner (I always find them), I also stopped by the Apollo. I had a root canal and hope for a crown next week. My dentist and her assistant are in the picture below. She is terrific, studied in the U.S. The only problem is the dentist chair is built for the average Indian and for me it's like trying to fit into an elementary school desk chair.

Now for the best part. The cost for the root canal and crown is 5,500 rupees, that's about $100. Time will tell if it's as good a deal as it appears.

Now for a story. One of the U.S. engineers had stomach problems about 6 weeks ago and not surprisingly, we thought it was the water. So did the doctors. After he didn't completely recover and passed out one day in the office, he had a CAT scan. The doctors found a cyst on his brain. This was a long-standing condition and had nothing to do with being in India. It was in fact the cause of the stomach problems and dizziness.

Since the diagnosis, he has had brain surgery to remove the cyst. Yes, that's right: BRAIN surgery. It was successful and he is recovering as this is being written. Fortunately, his wife has been here all along and his daughter arrived for a holiday about a month ago. Their support has been essential and how they all faced up to this and didn't come completely apart is a testament to the strength of the human spirit. The local staff has been wonderfully supportive, they really care and treat them as extended family. It's actually common for the whole family to accompany a sick relative to the hospital. It's always a busy place. This all makes the engineering difficulties seem easy.

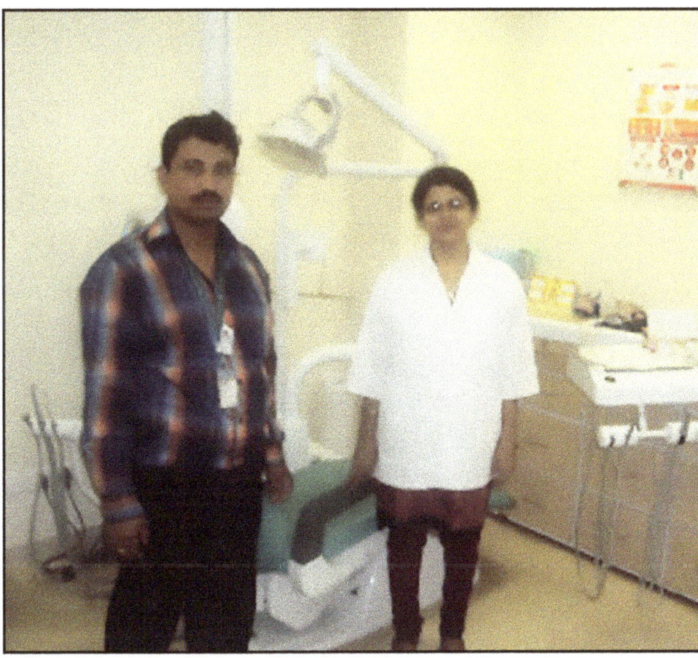

My dentist, Dr. Acharya, and her assistant Promod.

TUESDAY, JUNE 14, 2011

Monsoon

The monsoon arrived last week right on time. Last year there was little rain in eastern India, but this year is getting off to a good start. It has rained every day for a week and sometimes the thunderstorms in the late afternoon or at night are impressive. The main downside is the power outages which typically accompany them. The good news is that the heat has moderated with the overcast weather and breezes. There are no street drainage systems except on the main roads, so puddles and big ponds are everywhere. The soils are largely sandy, so that helps after the rain stops. You realize why it's best to live on what Indians call the first floor and what we call the second floor in the U.S.

Our street does OK because there is a little slope, but I'm glad we don't live at the end of the block. The first picture shows the house across the street, where three generations of a prosperous family live. We live on the second floor of the guest house. It has four bedrooms/baths, a kitchen, dining room, living room with TV, common area, and rooftop deck. The four bedrooms (I have the nicest one), are filled at the moment with the gang you see in the second picture, supported by our excellent cook and house boy (taking picture).

Even if we have a " wet monsoons" we will be high, dry, and well-fed on the second, I mean first, floor.

House across the street

Dinner with cook

SUNDAY, JUNE 19, 2011

I was just thinking...about India

Time: As Einstein said, relative
Rules: Followed to a fault officially, disregarded personally
Family: # 1.
Religion: Free form, the best
Culture: Soft and embracing
Women: Strong, spirit of the country
Men: Over 50 = Rigid. Under 50 = impatient and smart
Corruption: Same way all over the world, just more of it
Hospitality: Over the TOP
Planning: Thorough but uninspiring
Execution: Public sector poor, private sector good
Materials: Concrete and brick
Food: Spicy and good
Weather: HOT
Education: Overrated, big time
Politics: As Tip said, all politics are local; and here it's a democratic mud fight, but as Winston said, it's the best form we have.

THURSDAY, SEPTEMBER 15, 2011

Mumbai

Mumbai, still Bombay to some, is today the heart of a vibrant India. It's a cosmopolitan city, more than most in India, the financial capital and the gateway for the West to India. Speaking of gateways, the India Gate is shown below. It was built to welcome King Georges V in 1911 and the last British troops left through the Gate in 1947. Also shown is the Taj hotel, the site of the November 26, 2009 massacre launched from Pakistan that has further poisoned India-Pakistan relations, especially for the man-on-the-street. The hotel has been renovated and remains one of the best in the world. The chairman of Tata Industries, the hotel's owner, has done much to heal the damage, including financing the schooling of all the children whose parents were killed or injured in the raid.

Despite the cosmopolitan air, you can't escape the masses. The movie Slumdog Millionaire accurately depicts slum conditions, but on closer inspection most operate quite effectively given limited resources and no utilities. This is not to say that the plight of the ever-increasing number of urban poor isn't India's #1 problem. Yet right beside the slums are the best hotels, residences, and retail shopping you can find anywhere. I bought a colorful summer shirt which received favorable comments, which is normally not the case when I return from Brooks Brothers in Boston.

India Gate, Mumbai

Taj Hotel, Mumbai

SATURDAY, SEPTEMBER 17, 2011

Badminton

Indians love sports, singing, and dancing. In sports, after cricket and soccer, badminton is the game of choice. Chinmay has made a court in the park area in his residential complex with stripping, and lights for night play. It reminds me of the whiffle ball field I made as a kid in our backyard with foul lines, a "borrowed" home plate, and lights. In India it's best to play under the lights because the temperature and wind drop late in the evening. Wind is a real problem when playing badminton.

Since we have two local offices, one in Bhubaneswar and one in Cuttack, it was natural that one challenge the other to a match. The Cuttack team, which plays in the street in front of the office, was impressed with Chinmay's court. No chance of being run over by a car or cow. Unfortunately, they couldn't adjust to the plush surroundings and lost the match. A re-match in Cuttack is being planned, under the street lights. The game was played with lots of laughs and good sportsmanship but make no mistake, the execution was intense – and winning, with the attached bragging rights, was the only goal. Although the lights were fine for our play, they were a bit less intense than Fenway Park, so it took some time to adjust.

SUNDAY, SEPTEMBER 18, 2011

Flood

Last year the monsoon season, which lasts from June through September, was a bust in eastern India. In fact, there were drought conditions that affected crop yields. This year was a normal monsoon until 10 days ago, when the heavens opened for a week in the headwaters of the Mahanadi River. There is a huge flood control dam and reservoir in the mountains that got overwhelmed and when the floodgates (63 of them) were opened, a catastrophic flood roared toward the delta.

Much like New Orleans during Katrina, several flood levees failed and the rural farmlands in the delta flooded, driving thousands from their homes. The cities, which are also protected by levees that didn't fail, were in better shape. The cities were also lucky that the heavy rains were in the mountains, since the rain that falls directly on the cities needs to be pumped through the levees against the high river waters. The pumps, if they run at all, are too small to keep up with heavy rains, so less rain in the city was a blessing.

The flood has passed, but water remains in many areas and the cleanup and public health work, to provide potable water, food, and prevent disease, is ongoing and will be for some time. Also ongoing is a huge political argument about the operation of the big dam mentioned above. The dam has a combined flood control and irrigation mission and an operation protocol to accomplish the mission. However, during the monsoon, there is conflict between flood control and irrigation and the protocol allows for judgement in this regard. For example, if you keep the water level low in the reservoir to maximize flood storage, and the flood doesn't come, you are left with a half empty reservoir at the end of the monsoon season and not enough water to satisfy the irrigation needs in the dry season. If you do the opposite and the rains come with a vengeance, as they did this year, you don›t have enough flood storage and must release water at much higher rates, which creates downstream flooding. The opposition party is blaming the government for mismanaging the dam by allowing the water levels to be too high, minimizing the available flood storage when it was needed and thus creating a man-made flood. This debate will no doubt sell a lot of newspapers in the coming weeks, but will not help the people, and most are poor, who need government help NOW.

The first picture shows the water level on Monday September 12, one day after the high water mark, which was about 1 meter (3 feet) above what you see. The second picture was 5 days later, a clear improvement. The third picture shows shelters that the floodplain dwellers made along the road as they were driven out of their homes by high water. This was a common site all over the state of Orissa.

High water, monsoon

Low water

People displaced from flood plain during moonsoon

SATURDAY, SEPTEMBER 24, 2011

India Clients

One of the biggest problems dealing with public clients in India is the turnover of staff. The Indian bureaucracy has a habit of rotating people from post to post and even within one post, so the responsibilities can change frequently. Essentially it means you are getting a new boss just when you have figured out the old one. This is a problem, everywhere. Project management 101 teaches that you should not get a verbal promise from someone in authority, even if you trust him/her. Get it in writing. The person you trust may not be around, or have trouble with short-term memory, when you need the promise on paper. Even if you have written documentation, there can be complications, but you are in a better position to negotiate a reasonable outcome. An example will help to illustrate.

On one project we are designing sewers in an area with very flat topography. In these cases, there is always a trade-off between having a lot of deep sewers without pumping stations or keeping the sewers shallow and increasing the number of pumping stations. There is no right answer. Some engineers favor deep sewers over pumping stations (me) and some the other way around. Design criteria for our project were developed and accepted by the client that favored deeper sewers over pumping stations. The design was completed and was being reviewed, when the client's chief engineer changed. Guess what happened. Sure enough, the new guy liked pumping stations and didn't like deep sewers. More to the point he didn't like the previous chief engineer and would do almost anything to expose his "ill conceived" design. We were caught in the middle.

After much debate we changed the design and added a few pumping stations. Our costs were minor, since we had already tested most of the options. The bigger issue was the delay in the schedule. It wasn't the rework that took much time. It was the endless meetings needed to get a written decision directing us to make the changes, that took all the time.

The real issue will be a year from now when we are behind schedule and blamed for the delay. And guess what? There will be another chief engineer who will have no idea why the design took so long and probably hate pumping stations.

Life in the fast (?) lane.

SUNDAY SEPTEMBER 25

India Engineers

Much has been written about the number and quality of engineers coming out of places like India and China. I can only speak about India, and in India I make three general engineering classifications. First would be the software/IT engineers, second the engineers in the manufacturing sector, and third the engineers in the buildings and infrastructure sectors. I'm most familiar with the latter.

First, contrary to popular belief there are not enough engineers of all types to serve local needs, to say nothing of exporting engineers outside India. This situation is getting worst especially for the manufacturing, building, and infrastructure sectors. I read recently that artificial intelligence programs will be taking over much of network maintenance, so be ready you IT engineers.

There is no professional registration process in India like the U.S., so you are an engineer by virtue of your schooling. Your degrees and where you got them are the coin of the realm. The seven India Institutes of Technology (IIT) are as good as MIT and the graduates are top notch, but from there the quality can drop off quickly. There is a clear recognition that more graduate engineers are needed, but the education system is struggling to expand with quality.

As I have said before, there is a different mindset between the older engineers and the younger ones. The former came up the hard way and follow time-honored standards, even as time has passed them by. For example, the water/wastewater design Bible was issued in 1994 and as you might imagine is a bit out of date but still followed with reverence. It's finally being updated; however, the project has a technical advisory board of 30 respected engineers, which means they are all old. Since, in my experience, it's difficult to get three engineers to agree, 30 will be a real mud fight. I bet the new manual will take another 10 years to publish.

The younger engineers are bright and eager, but don't have the cadre of seasoned professionals to learn from. Also, in buildings and infrastructure, the contractors run the show, not the engineers. Except for a few experienced contractors, many do not have high standards, and the engineers tend to design to the "lowest common contractor." Most projects only get enough design (25%) to be incorporated into documents that form the basis for the contractor's bids. The low bid contractor gets the job, finishes the design, and builds the project, with painfully little oversight. This method denies the engineer the opportunity to stay involved through construction, and consequently the quality of the final product suffers. One needs only to look around at the infrastructure and buildings to see the results of low bid work.

You might have noted that I'm laying the blame for poor quality on the contractor, not the engineer. Surprised?

FRIDAY, SEPTEMBER 30, 2011

The Fish Market

I was complaining the other day about the lack of fish and its poor quality given that we are only 40 km from the Bay of Bengal and located in the delta of the huge Mahanadi River. The fish are small, freshwater, over cooked, and bony. Before I could finish, my good friend Kalyan who is very proud of his home state of Orissa, offered to take me to the main fish market in Bhubaneswar and buy any fish I wanted.

So the next morning we went, and I saw the variety of both fresh and saltwater fish available locally. Most are dead, but some are kept in buckets alive in case you are interested in freshness. The preparation is done while you wait by the many fish sellers. As you can see each man squats in front of his catch and prepares what you want on a huge, sharpened blade protruding from his squatting block. It can get quite messy with heads, tails, fins, and blood scattered everywhere. The pictures show the more tidy spaces.

After much discussion, and some uncertainty about the fish, we played it safe and brought shellfish. You can see in the photo the variety and high quality of the prawns (shrimp) and soft shell lobsters. We also now know what is available and can instruct our cook to add more fish to the menu.

A corner of the fish market

The final selection, shellfish

MONDAY, OCTOBER 10, 2011

Chinese Fishing Nets

In south India's tidal backwaters along the Arabian Sea, you find a one of a kind fishing technique. Brought to India by the Chinese centuries ago and left behind, it's the lazy man's way of fishing, once you get the whole contraption built. You can't find this type of fishing in China any longer, they have graduated to big ships and nets.

There are several designs, but the simplest one has two long wooden poles (arms) that hold a net open in the water. The arms are attached to a pivot point on the dock and when they are lowered, the net goes underwater and traps the fish swimming by. The net is pulled out of the water by ropes that pass over a second set of poles at about 90 degrees to the first set, also attached to the pivot. This is all aided by a counted weight system of ropes and rocks. Once balanced correctly, one person can pull the net up and two people makes it quite easy. I tried.

The wise fishermen know what time to fish (day or night) and how to tilt the net to favor tidal flows. There is an internal net that is fashioned to allow the fish in but not out. Crabs can also be part of the catch

The first picture shows the net up, and the second shows the catch. Only one this time. I didn't bring good luck, but I did bring a few rupees.

Net up

The catch (disappointing) and my guide

TUESDAY, OCTOBER 11, 2011

Puja

In eastern India, the biggest annual festival is the Durga Puja. It's the worship of the goddess Durga for her victory over the buffalo-headed demon Mahishasura. Essentially good beats evil. It's a 10-day festival, but things don't get started until day 6 or 7. Temporary shrines are constructed which house images of Durga for the worship part. These are huge and in some cities, copies of iconic structures like the Statue of Liberty and the Eiffel Tower are also built. The best part is the carnival around the shrines. It's a little like the Fryberg fair, but the cows are kept out. There are Ferris wheels, rides, games, and much food. Great family entertainment.

Kalyan took me around. We first stopped to pay our respects to Durga, made a small offering, and got a good luck string tied around our wrists. Next, we walked through the carnival. The kids were having a great time, and the food smells were tempting, but I limited my sampling to some nuts and lived to tell the story.

On the last night all the statues of Durga are put on carts or flatbed trucks for a procession (parade) to the river for immersion. This starts at dusk and can take much of the night, to everyone's great delight. Nothing is reused, it all gets washed down the river. The next day, after rising late, the plans begin for next year.

Durga in all her splendor

Carnival with Ferris wheel

MONDAY, OCTOBER 17, 2011

Gate Ball

Gate Ball is another India sport played with great passion, but not as much energy as cricket, football (soccer), or badminton. This becomes clear when you find out that one of the top players is an 85-year-old great grandmother and also when you realize that this is the Asian version of croquet. Actually, it was played once in the 1900 Olympics, in an earlier form started in France.

The current game originated in Japan after WWII and now there is international competition among many Asian and near Asian countries. Each country has teams and elimination tournaments, with the winner going to the annual international games in January. We played with last year's champs as they were training for their next tournament. You can work up a sweat, but only because of the temperature and humidity.

The field of well-trimmed grass is only about 15 by 25 meters (50 by 75 feet). There are 3 gates and one central pole that is the finish point. Each player starts by knocking his ball through the first gate. If you don't get through, you go to the end of the line of 10 players. Each team has 5 players, one team has odd numbers and the other even. See the photo. The idea is to go through each gate (1 point) and finish by hitting the central pole (2 points). The team with the most points after 30 minutes wins. That's the simple explanation, but like much of life, it's all about strategy.

Like croquet, if you hit one of the other team's balls, you get to hit it out of bounds. This, together with the rotation of players, makes for some interesting games. There are periods where it looks to the uninitiated that players are hitting their balls in the wrong direction and for unexplained distances. Then the trap is sprung, and one player goes around the field scoring points and hitting the opposition balls out of bounds. At this point there is a bit of excitement, although not quite like the final minutes of a close Patriots game.

Take my word for it, because I don't think this is coming to the U.S. anytime soon to challenge the NFL on Sunday afternoons.

Right through the gate

Sending the opposition's ball out of bounds

TUESDAY, NOVEMBER 1, 2011

Dogs

It's hard to walk down a street in any city in India and not see stray dogs roaming around in packs of 3 to 6 or more. They are normally medium size, in the range of 30-40 pounds and would weight more if they had a normal diet. Their coats are shabby due to poor living conditio\

The dogs are usually doing one of the following, ranked by time. Sleeping, standing with a blank stare, looking for food in a garbage heap, or a distance forth, fornicating. See the pictures below. This is a family blog, so no fornicating.

During daylight you don't give them much thought, and they generally mind their own business. Occasionally one gets pissed off for some reason and snarls, which gets your attention. For this reason, I keep my distance. At night it's another story. I have been warned several times not to be in a deserted area after dark, not because of hoodlums, but because the dogs run in packs and can attack you. The daylight snarls give meaning to this warning. Fortunately, with so many people, there aren't too many deserted areas until well after midnight.

You don't see many stray cats in India, and I don't know if this is because of the dogs or some other reason. Certainty the dogs don't follow the vegetarian habits of the human population. After observing a few garbage dumps, it's clear the dogs are carnivores.

The Gang

Garbage patrol

MONDAY, NOVEMBER 14, 2011

Indian Picnic

I was invited to a picnic hosted by one of the senior bureaucrats in the municipal government. This was in October, so I'm a bit late with the post.

Many upper-class families live in the city and also have agricultural land holdings. In this case, the family has about 40 acres of land less than a one hour drive outside of town. The land is served by canals, but more important this one has wells that supply irrigation water year round. This allows for abundant fruits and vegetables and three crops of rice annually. The rice is sold, but everything else is for family use. They have machines to plant and harvest the rice, but this is a recent addition. The work is done by two tenant families and the owner's extended family use the farm almost every week as a summer home.

There were about 40 folks at the picnic, which occurs several times a year around a small reunion of the owner's college classmates. One of the women I met got her PhD at the University of Maryland, not long after I was there, so we had a fine chat about that.

The Indians LOVE to play games, so we had some fun. I actually won a form of Bingo called 'Hossey' and was appointed to give out the prizes as the honored guest. There were spelling and memory games in English and since I had no chance of winning, I lucked out with Hossey, which requires absolutely no skill other than recording numbers, providing you can understand the Indian English.Compared to what I'm used to, the food was just OK, but they were kind enough to have a fork just for me.

Farmhouse

Rice to the left and fruit trees to the right

The women's social circle, always separate

The prize winners

THURSDAY NOVEMBER 24, 2011

Native Dance

We visited Raghurajpur, an artist village that is famous for patachitra, a cotton cloth coated with a mixture of gum and chalk and then polished. The artist then marks the surface in detail with pen and brush, with bright colors. About 100 private homes have small workshops where the crafts are made and sold.

Also in the village is a small school that teaches a native dance done by young boys. Before I was told they were boys, I thought they were girls, given their dress and makeup. The dances, including yoga positions and movements, were impressive. Much twisting, turning, and tying in knots in ways seemingly impossible for the human body. They also used a technique of slapping their bare feet on the marble floor that gives off a "fire cracker" like sound and makes the dances all the more exciting. The dancers were accompanied by two musicians, unfortunately out of the picture below.

The dancers

THURSDAY JANUARY 26, 2012

Mumbai

Pete and Jill (niece of Pete and Rick) are on a visit to India and the first stop was Mumbai. We took a harbor boat ride to Elephant Island, a world heritage site with massive cave/temple and a world class sculpture of Lord Shiva. The only elephant was made of stone, but there were plenty of monkeys. The harbor is very commercial with silty water, but remarkably free of garbage.

The boat leaves from the Gateway of India next to the Taj Mahal Palace Hotel. After our visit to the island, we had a pleasant late lunch at the Taj. Lap of luxury. We than went to the Crawford Market, from British times, where you can buy fruits, vegetables, and a lot more if you can stand the sights. We smelled many spices and bought some curry to bring to our friends in Delhi.

Pete and Jill at the cave/ temple, Elephant Island

Taj Hotel and India Gate

SATURDAY, JANUARY 28, 2010

Jaipur

Our next stop was Jaipur, which is the capital of Rajasthan, the largest state in India west of Delhi. It's called the Pink City because the local Maharaja required the citizens to paint the buildings pink in honor of a visit by the Prince of Wales in the 19th century. Not much pink left today, but the population is now over 3 million people. It's India at its chaotic best and worst.

The downtown has some fine palaces built in the last 200 years, but the best sights are the 3 forts on the hills outside the city. The main fort and old capital, really a palace, is on a spectacular hilltop setting. It also gives you some peace and space after navigating the crowded downtown, which is a bit intense.

We were also lucky to be staying at a hotel in a converted palace of one of the lesser Maharajas. This too was an oases from the din of city life. One night on the cricket field the hotel organized a game of elephant polo. Not exactly a fast game, but one that is hard to stop once it gets started. Quite fun.

The Jaipur Literature Festival was on when we were there. You might have read about Salman Rushdie being banned from the festival because the authorities didn't want to offend the Muslims, with elections coming up next month in several nearby states. So it goes in the biggest democracy in the world. At least the controversy sold a ton of newspapers.

Jill and Pete from fort with Summer Palace in background.

Elephant polo

SATURDAY, FEBRUARY 4, 2012

Last Stop

After a visit to the Taj Mahal, without me, Jill and Pete's last stop was Bhubaneswar. They were good sports and stayed in the Guest House with me. This was an adjustment after over a week in Taj Hotels. Just for reference, the Taj Hotel chain is the same one that took over the "old" Ritz on Arlington Street in Boston.

They both appreciated the local atmosphere, a bit off the tourist trail even though Bhubaneswar is called the city of 1,000 temples. They only saw a few, but did meet the local staff and got to understand the work we are doing firsthand. We also visited a slum that had received funding for USAID to install outdoor water well points and latrines. This is a huge improvement that makes water more assessable and reduces open defecation. Since there is a religiously significant pond right next door, the water quality will improve, to say nothing of public health. Your tax dollars at work.

There are about 300 defined slums in Bhubaneswar with several hundred thousand people. These are functioning neighborhoods where the government is making efforts to improve living conditions, a worthy and immense task. This doesn't include the homeless population, which is another story.

We got to the countryside and Chilika Lake, took a boat ride, saw the dolphins, and walked across the sandbar separating the lake from the Bay of Bengal, "where the dawn comes up like thunder out of China cross the Bay".

Pete making friends

Jill against the Bay of Bengal

TUESDAY, FEBRUARY 14, 2012

The Three Idiots

Don't misunderstand the title. It was a movie several years ago about three guys going to college based on their parents' expectation that they all become engineers. As you might guess they weren't at all interested in engineering and eventually took different paths, much to their parents' dismay. They were anything but idiots. It's a comedy with biting commentary on parents' misguided and overbearing expectations, and hit a nerve in India. Sound familiar?

The photo below shows my three "idiots." I spend much of my time with them, rely on them, trust them, and couldn't get by without them. Number one is Chinmay, second from left. His wife is on the right without their baby girl, home with grandparents. He is a local boy, but don't let that fool you. He is as sharp as a tack and has diplomatic talents that would be envied in Foggy Bottom. Excellent technically, with a master's degree in environmental science, his vast talents prevent him from doing the technical stuff. He is far more important getting the technical message out in a way that clients and the public can understand. He also keeps client relations on an even keel, anticipates problems, and always has a solution. Normally he's right, and very sophisticated. In short, he's my secret weapon on whom I rely every day. He will go far in any country and in any environment. I hope he can visit the U.S. in the future, so you can meet him and see what I mean.

Next is Amod, in the middle, also with his wife and lovely daughter. Amod has been the go-to technical guy on all sorts of issues. He has a good feel for construction issues, which I sorely lack, so this has been of great help to me. Amod is from Pune, 1,000 mile away, where he and his wife's families are located. Family being so important, he made the difficult decision to return home, with my blessing. I'll miss him both professionally and personally. In a previous life he was the lead singer in a band and I will miss hearing him sing, which usually took a few beers and some arm twisting. Worth it!

Finally is Mr. Otaguro, our Japanese connection. Since we do so much work funded by Japan, a direct connection is necessary. His wife and son live in Chicago and we help him with visa issues, so his family can stay there. He has many years of experience navigating the Japanese funding process all over the world, so there isn't much he hasn't seen. Mr. Otaguro is also our contracts guy, another area I'm not terribly fond of, and he can prepare a wicked financial spreadsheet. Since we are both ex-pats we have a lot of meals together. He is good company and never complains about the small stuff. What we call low maintenance, and given all the other things to contend with, this is a blessing.

My hat's off to all three and their wives. They are clearly not idiots!

Mr. Otaguro, Chinmay, Amod, Amod's wife and child, me, and Chinmay's wife (baby is home)

FRIDAY, FEBRUARY 17, 2012

The Brahmaputra in Assam

The state of Assam is the far eastern portion of India on the other side of Bangladesh, next to Burma (Myanmar). We won two projects funded by the Asian Development Bank (ADB) and Satish and I visited Guwahati, the capital city, last week to sign the first contract. One contract is for infrastructure work and the second is for flood contro

After the signing we had dinner with the commissioner of the Guwahati Municipal Corporation, the chief bureaucrat and our client. The folks who hold these positions are the top professional bureaucrats who move around the country and are really sharp. Too bad the lower level folks are far less competent. I told him I had heard a lot about the Brahmaputra River and wanted to see it the next day. On the spot he called an assistant to set up a short trip to a temple on one of the islands in the river.

The Brahmaputra comes out of China through the Himalayas into a broad valley that runs the full length of the Assam. Guwahati sits in the valley on one bank of the river, which is miles wide and there are no bridges. Crossing requires a boat, but there isn't much but forest on the other side. The Brahmaputra joins the Ganges in Bangladesh to form the largest delta in the world, essentially the whole of Bangladesh. The River is about 2,000 miles long, about the length of the Mississippi but with a larger flow due mostly to the monsoon rains and snow melt from the mountains. It has about the highest flow per square mile of any river in the world.

Needless to say, there are serious floods. Part of our work will be to improve the levee system (dikes) and help institute better floodplain management. Putting homes on stilts, if one must live in the floodplain, is one solution. Upstream dams are another, but for many reasons that isn't going to happen soon.

Our trip to the island started early before the public opening at 10 am, thanks to the commissioner. The whole trip, with time to worship, took less than 2 hours. It was a wonderful experience seeing the river and Hindu rituals firsthand.

The island with the top of the Temple between the trees

Worship in the inner sanctuary

Morning bath. Monsoon floods rise into the trees and beyond

SATURDAY, FEBRUARY 25, 2012

Temple on a Hill

In Hindu mythology, Lord Shiva, the Destroyer, got a bit upset when he found the charred, lifeless body of his new wife Sati. Not surprisingly, his response was to begin to destroy the earth. Fortunately, Lord Vishnu, the Preserver, stopped the destruction, but in the process the body of poor Sati was torn into 51 pieces and landed all across India. Upon these 51 widely scattered places, temples were built and today most are World Heritage Sites. One, where her yoni (genitalia) fell, was on a hill just outside Guwahati overlooking the Brahmaputra River. It the largest most vibrant temple complex I have visited, crawling with priests.

The main ritual centers around the sacrifice of goats or other available animals. The dominant color in the temples and the surrounding village is red, which apparently helps soften the visual impact of the spattered blood. There is some pushback about the sacrifices from the younger more modern worshipers, so they are less frequent, and we (fortunately) missed them.

Unfortunately, the sky was hazy the day I arrived limiting the views, but the people were not diminished in their enthusiasm, especially the kids -- as is the case all across India.

Happy girls, serious boys with white guy

The priests in conference. 'Lunch before or after killing the goat?'

SUNDAY, FEBRUARY 2 2012

Lunch on the Beach

Kalyan is our advisor on all things Orissa. Together with Chinmay they make a formidable team, one I couldn't do without. Kalyan took me on a drive to the beach one Sunday to have lunch and show me a terrific vacation spot that has about 20 well-equipped cottages within 100-200 feet of the beach. A place for a private, peaceful getaway, or just to hang out, catch some sun, snorkel, and walk the beach.

The beach is at the mouth of a small inlet/river that enters the Bay of Bengal at an angle. The inside of the inlet is the beach, and the outside is a sandbar that protects the beach and eventually disappears as the waters mix with the Bay. As a result, there is no surf on the beach, so you need to swim across the inlet and walk cross the sandbar to get to the Bay.

Alternatively you can ride the outgoing tide and get washed into the surf. Given rip tides in the Bay, I chose to restrict my swim to the inlet. The water wasn't warm (Florida) or cold (Maine), more like the Jersey Shore in the early summer. Refreshing.

We had a 4 course meal of fish, fish, fish, and ice cream. There is a chicken flu (like swine flu) in this part of the world and you can't get chicken. Good thing. The first course was large local prawns (shrimp). The second course was a whole small local fish and the third course was small fillets of larger local fish, advertised as boneless. At least the ice cream had no bones and we had an ample supply of Kingfisher beer to wash them down.

The lunch table on the beach. Second course

The beach

SATURDAY, MAY 12, 2012

Getting Paid

Without a doubt, the three biggest business frustrations in India are: getting paid, getting paid and getting paid. In almost every case it has nothing to do with problems related to schedule, budget, or general client satisfaction. The client is your best friend and satisfaction prevails. So why does it take so long to get paid? The day-to-day reasons are endless, but the problems generally fall into two categories, one which follows the other.

First relates to the contract terms. Too many of our contracts with local and state governments have payment terms that are tied to milestones. There may be an advanced payment (10%), followed by another

20% with a deliverable (topography survey), and so on. The problem is not in the beginning, it's toward the end when the last 20 to 30% payment is tied to an approval of some sort, normally of an engineering design and/or cost estimate. Since there can be multiple approvals by your client and higher levels of government, your ability to control the process is limited. Time passes, questions continue, and revisions are added to revisions. Finally you think you are done and so much time has passed that the client wants you to update the design, or more likely the cost estimate. The problem is you can't even submit the last invoices for 20 or 30% of the project costs until the approvals are official, which means in writing. Getting the written approval even after everyone is happy is another story. Let us be optimistic and believe we can get the approvals in a reasonable time (really optimistic). This brings us to the second stage of the process: actually getting paid. Of course this stage applies to every invoice, not just the last one.

Anyone who has tried to get paid from a public agency in the U.S. understands the frustration. Think about submitting an 80 page invoice to the MBTA for $8,000 of work. The same principles apply in India, only worse. In one case we needed 9 signatures by 5 different people: four signatures going up the chain of command and five signatures going back down the same chain. Each of the signatures is accompanied by a handwritten note in the file. To get this done you need to be actively engaged at each step to make sure the file doesn't go to the bottom of the pile (see photo below). We have gotten one client to eliminate going down the chain of command, eliminating 4 steps. A small victory.

As you might imagine, there are times when someone in the chain has a problem. Most can be fixed with an explanation or additional back-up information, but some are just problems. In these cases the person in question would like you to say something like, "what will it take to make the problem go away?" Since we can't say this and provide what is normally expected, the process continues and frustration mounts. So what is the solution? My short-term solution is usually to scream. Long term, we need to avoid work with local and state governments in favor of work that is not milestone-based with the central government and work that is funded by donor agencies, like the Asian Development Bank.

You don't want your invoice to be at the bottom of one of these piles

SUNDAY, MAY 20, 2012

Safari

I just finished a South African safari and was not disappointed with the trip. The Big 5 comprise the most dangerous animals throughout Africa: elephant, lion, leopard, rhino, and water buffalo. It should be the Big 6 because I don't know how the hippos got left off the list. They actually kill more people than any of the others and are very aggressive. However, if you ask a guide which is the most dangerous, they will say a herd of water buffalo.

We had a chance to see them all at close range from a land rover. One day I went on foot with our guide tracking an elephant, rhino, and water buffalo. This gets the blood flowing, especially when a rhino is looking directly at you even though you know they have poor eyesight. Their ears move independently trying to hear you and they have an excellent sense of smell. For the rhino and the elephant (smells with her trunk), you need to keep up wind and stay still -- easier said than done. With the water buffalo, you stay close to the land rover because their typical pattern is to run away and then turn and attack. At which point you run. We didn't get too far from the rover and fortunately the buffalo didn't turn after we spooked them. To give some comfort, the guide carries a rifle and has very impressive training for this type of work.

Our guide, William, was half black and half white, which is not a good place to be in the social structure. He was a great guy, very knowledgeable, and gave you a sense of security where there wasn't much. The two pictures show a contented lion having just feasted on a wildebeest and a lady elephant. Say what you want about the king of the beasts, but size matters. You can't beat seeing elephants, rhinos, and hippos up close.

King of the beasts

Big mama

MONDAY, MAY 21, 2012

Victoria Falls

After the safari, we went to Victoria Falls which is on the Zambezi River between Zambia and Zimbabwe. Lots of Zs. We stayed in Zimbabwe, which is emerging from a bankrupt economy and trying to get back on its feet. There is essentially no currency, so you use dollars. The kids on the street sell 1 billion Zimbabwe notes for 25 cents as souvenirs. Zambia is doing better.

The Falls are amazing at twice the height of Niagara and 50% longer. We were there at high water so the mist was impressive and signaled the Falls presence miles away. We took a sunset cruise above the Falls where the river is wide and home to elephants and hippos. Watching them for extended periods is better than the sunset, which was quite brilliant.

This section of the river was used 100 years ago to land seaplanes on the trip between Cairo and Cape Town.

We took a helicopter ride to get a view of the Falls and were not disappointed. The river below the Falls, unlike upstream, flows in a great zig zag cataract similar to portions of the Colorado River. The water was too high to do rafting, which along with bungee jumping off the bridge, are big tourist attractions. Despite much encouragement from the locals, we only watched the jumping.

Falls with bridge just downstream

Bungee jumping from the bridge into a rainbow.

TUESDAY, MAY 22, 2012

Qatar

I stopped in Qatar for several days on my way back to India to help develop a JV between Tetra Tech and a local firm. In order to do work in Qatar, or in most Middle Eastern countries, you need to be associated with a local firm. The firm can simply serve as your agent representing you in the country or can play a meaningful role. In our case our partner owns a construction company, among other businesses, and will therefore do lots of real work. This is a good thing. I tried this several years ago, when Tetra Tech had no interest. Now they do.

As most know, Qatar has lots of gas (the good kind) and is building like crazy (Dubai?) to get ready for the 2022 soccer World Cup. For these reasons and others there are a lot of opportunities in Qatar. Since there are only several hundred thousand nationals who are either filthy rich or supported of by a generous government, most of the work is done by a million or two ex-pats. Many come from India, the Philippines, and other parts of Asia.

I don't begrudge the nationals taking full advantage of their good fortune. Qatar is one of the most stable and progressive Middle Eastern countries. Al Jazeera is broadcast from Doha, the capital, and the ruling family is trying to use their wealth to prepare for the future, whatever that will be.

However, looking at the massive skyline in a traffic jam with no people in sight is a strange feeling. Especially so when compared to the beehive of people that is India or even Boston. A lot had to do with the season, summer, when temperatures are well above 100 in the shade. People (men) do come out to cafes in the nicer weather to smoke water pipes and gossip late into the night. Of course even in the winter, it is a desert remember, almost everything is inside. The Pearl is a man-made island community with a mix of office, retail, and residential for the well-heeled. It's about half completed. We went into the shopping area with the most expensive retailers in the world and virtually no one was there. I suspect a customer typically drops several hundred thousand dollars per visit, instead of just several hundred dollars. This makes the retail work. I passed the Ferrari dealership at the end of the shopping center, but didn't stop in.

The skyline from across the bay. Those are 30-40 story buildings

Inside the shopping center at the Pearl

WEDNESDAY, OCTOBER 10, 2012

Summary

The past two and a half years have been quite a ride. On a personal note, I couldn't have asked for a more interesting experience meeting wonderful people and beginning to understand different cultures. People like Satish, Poonam, Chinmay, Kalyan, Amod, and Otaguro stand out among many, as people who made me feel at home and supported me unconditionally. Of course Satish is closest to my heart as he literally made me part of his family. His wonderful daughter and son call me "uncle".

On the professional and business side, things were less positive. We did significantly improve our qualifications by winning and working on large water-related projects. This allowed us to move from being a subcontractor to being the prime, because in the beginning we didn't have the minimum qualifications to be the prime. No company will ever be successful as a subcontractor in the long run. You need to be the Prime, so you are in control, at least as much as you can with international work. In many ways, unlike with U.S. clients, winning work is the easy part. Staffing and contractual issues were a constant struggle, but above all else getting paid was the killer.

As I have said many times, India is heading in the right direction and has made huge progress in the last 10 years on developing a business climate that is transparent and fair. This is a little like President Obama's re-election argument. "We started in the cellar, have made great progress, but there are still miles to go." This may be true, but the current situation still stinks in India. Likewise I can't justify doing business in the current business climate in India, with a few exceptions.

Finally, to do business in a foreign country you need to be present. Relationships are the foundation for success and they can't be done remotely. This of course required me to be away from family and friends far too much. I initially expected the serious time in India to last 6 months, but it stretched into years. Now enough is enough and my place is home.

Happy Trails.

BLOG INDEX